Greener Products

Sustainability and its competitive advantage are the goals of every company and any brand that wants to stay successful in the marketplace. Customers gravitate to brands that manage sustainability issues well. *Greener Products: The Making and Marketing of Sustainable Brands*, written by a renowned sustainability expert, continues to address the latest developments in the extremely fast-moving field of sustainability. The third edition is thoroughly updated, introduces new case studies, and includes a new chapter on green marketing. With over 40 case studies, it explores the best practices of leading global companies and helps readers learn what it is that makes them successful.

New in the Third Edition:

- Presents, in a practical way, the best practices of sustainable brands in a global economy.
- Addresses the most current sustainability topics like circular economy, plastics in the environment, biodiversity, climate change, green chemistry, etc.
- Includes current marketing information on consumer trends to purchase greener products.
- Incorporates the latest pressures on companies to address sustainability, retailer programs, business-to-business expectations, ESG raters, rankers, and stock funds.
- Covers best practices of companies from various industries on how to make and market greener products.
- Provides current tools for making products more sustainable and methods on how to market sustainable improvements.

This book serves senior undergraduate and graduate students in programs focused on sustainability, as well as academics and corporate sustainability leaders. The previous versions have been used to teach courses on sustainability, product improvement, introduction to sustainability, green marketing and sustainability, and sustainability policy. Any university that teaches a course on sustainability and any company or individual interested in making and marketing more sustainable products would benefit from the new edition of this book.

Greener Products
The Making and Marketing of Sustainable Brands

Third Edition

Al Iannuzzi

CRC Press
Taylor & Francis Group
Boca Raton London New York

CRC Press is an imprint of the
Taylor & Francis Group, an **informa** business

Designed cover image: © From previous Edition

Third edition published 2025
by CRC Press
2385 NW Executive Center Drive, Suite 320, Boca Raton FL 33431

and by CRC Press
4 Park Square, Milton Park, Abingdon, Oxon, OX14 4RN

CRC Press is an imprint of Taylor & Francis Group, LLC

© 2025 Al Iannuzzi

First edition published by CRC Press 2012

Library of Congress Cataloging-in-Publication Data
Names: Iannuzzi, Al, author.
Title: Greener products : the making and marketing of sustainable brands /
Al Iannuzzi.
Description: Third edition. I Boca Raton, FL : CRC Press, 2025. I Includes
bibliographical references and index.
Identifiers: LCCN 2024007881 (print) I LCCN 2024007882 (ebook) I ISBN
9781032579870 (hbk) I ISBN 9781032579856 (pbk) I ISBN 9781003441939
(ebk)
Subjects: LCSH: Green marketing. I Green products.
Classification: LCC HF5413 .I16 2025 (print) I LCC HF5413 (ebook) I DDC
658.8/02--dc23/eng/20240514
LC record available at https://lccn.loc.gov/2024007881
LC ebook record available at https://lccn.loc.gov/2024007882

ISBN: 978-1-032-57987-0 (hbk)
ISBN: 978-1-032-57985-6 (pbk)
ISBN: 978-1-003-44193-9 (ebk)

DOI: 10.1201/9781003441939

Typeset in Times
by KnowledgeWorks Global Ltd.

Contents

v

Foreword

Despite a landscape of political and social tension, the need for reinvention of our economic engine around the world in support of greener product design has never been more pressing – nor has this need been more widely recognized. The world has woken up to the realities of climate change driven weather disruptions that are resulting in a record-breaking number of multi-billion-dollar disasters around the world. Devastating fires, water shortages, biodiversity loss, and even pandemics all create new risks to supply chain security, economic stability, and the predictability necessary to confidentially succeed at running a business today. The continuing drawdown of the natural resources we depend on to power our economy points to two possible paths – strip our planet bare and experience the devastation to humanity of having done so, or learn to redesign the way we make and do everything – starting with the way we design and deliver products to market.

The good news is that in this mandate lies opportunity for brands and business as well. More and more of us are experiencing firsthand the implications to ourselves, our families, our livelihoods, and our communities of an unchecked deterioration of our environmental eco-systems. As a result, consumers have higher expectations than ever of the companies they buy from. According to our own **Sustainable Brands Socio-Cultural Trend Tracker™**, fielded in partnership with IPSOS (*Institut Public de Sondage d'Opinion Secteur*) in August of 2023, between 69% and 80% of consumers depending on generation believe that companies are not doing enough in terms of supporting sustainability and ethical best practices. When it comes to the younger generations – Gen Z and Millennials 72% and 69%, respectively, report being willing to pay more for products and services that are ethically and sustainably made; 75% and 79%, respectively, say they would switch brands if a competitor offered a more sustainable version of the same product. And all of this is resulting in positive outcomes at the check stand. According to the **Sustainable Market Share Index™** study fielded in 2022 by NYU Stern Center for Sustainable Business, **products marketed as sustainable are responsible for nearly *a third of the growth* in consumer packaged goods (CPGs) from 2013 to 2022, and market share growth continues year over year.**

- Products marketed as sustainable now hold a 17.3% market share, up +0.3 percentage points versus 2021, continuing to grow in the face of high inflationary environment.
- Products marketed as sustainable grew ~2× faster than products not marketed as sustainable and achieved a 5-YR compound annual growth rate (CAGR) of 9.43% vs. 4.98% for its conventional counterparts.

Those brands which honestly seek to understand and improve on the direct and indirect impact they have on all their stakeholders—not just their customers and shareholders—will be the ones that take the lead and stand the test of time in the 21st century. Brands don't have to be perfect, but they *do* need to be authentic and transparent about their impacts, and they *do* need to show continuing improvements in the way they are addressing their environmental and social impact. We have a long way to go to get to a fully sustainable economy—and it will take a commitment from all parts of the economic system, including consumers, to get there. However, the degree to which a company aligns a meaningful brand promise with the way it designs, builds, and delivers its brand to market is key. And business leaders *do* have an opportunity to drive sales and build customer loyalty and brand value by positively addressing the social and environmental challenges we face through purpose-driven product innovation.

Building upon his first two editions of *Greener Products*, Al Iannuzzi and his contributors make the case for why bringing more socially and environmentally smart brands to market is today's opportunity and tomorrow's table stakes. Through reviewing some of the many case studies of leading companies' greener product development and marketing programs, readers can glean best practices of smart business strategy and sustainable innovation. *Greener Products* offers more validation that sustainability doesn't have to come at the sacrifice of profit.

The future of business lies at the intersection of what the world needs, what consumers want and what your brand equity gives you permission to deliver. If you want to know how to make your brand more sustainable and how to sell your better brand to your customers, *Greener Products* explains the details. As you'll experience from reading this book, companies that use sustainability as a driver of innovation and then smartly market their benefits will win. As Al says, when you "hit the *sweet spot*, of having a truly greener product that is communicated in an appropriate way, *everyone wins*."

KoAnn Vikoren Skrzyniarz
Founder/CEO
Sustainable Brands Worldwide

Preface

The world's resources are under more pressure than ever before. The global population continues to grow and more importantly a rapidly expanding middle class in developing nations such as China and India. With increased wealth, there is a desire to have products that more developed nations have: cars, computers, mobile phones, televisions, iPads, etc. More of these products means more of a demand for raw materials, increased greenhouse gases, more minerals extracted, and more waste to be disposed of. But the even bigger question is: Where will all of these raw materials come from? Regardless of how you feel things are regarding environmental pollution and the future of the world, we all can agree that there is going to be more competition over raw materials to make the products that the growing middle class desires.

At the very least, competition for resources will make products much more expensive to make and, in a worst-case scenario, would severely restrict our ability to provide these products due to a lack of raw materials. Considering this forecast, business as usual is not an option. Institutionalizing sustainability in new product-development processes is an imperative. Businesses must respond and do things differently, evaluate life cycles and minimize use of raw materials, and consider impacts during customer use and product end of life. Deploying such concepts as design for the environment, green chemistry, and biomimicry needs to be commonplace and built into the way new products are brought to market. This is the reason I authored this book; to offer guidance on how to make products greener or more sustainable and how to appropriately market them.

Before the first edition of *Greener Products*, I was unaware of any books that combined the concepts of design for the environment/industrial ecology and green marketing into one book, and I believe that this third dition of Greener Products is still the only one that combines these concepts. Whenever I speak about greener products, there are two things I usually say:

1. *There is no such thing as a green product.*
2. *What good is a greener product if no one knows about it?*

The reason for these statements is that life-cycle assessments have shown that every product has impacts, from raw materials to transportation, manufacturing, customer use, and end of life. Every product can be improved in some way, which is why I use the term Green*er*. Once you have a greener product, you must appropriately inform your customers why it is greener; if you don't do this correctly, it is essentially worthless to have an improved product since no one will know about it. That's why it is critical to *make **and market** greener products*—**the communication aspect is just as important as having a greener product**.

The words we use are important, and in this book, I use the terms greener products and more sustainable products synonymously. These terms are descriptors for products that demonstrate improvements in environmental or social areas. I should also mention that **I don't believe that any corporation is *sustainable*** or is *a sustainable business.* In my mind, **sustainability is a journey** and to the point above, every product and every corporation have an impact. There are a lot of things we can do to minimize environment impact, but we can't eliminate impacts completely, that's why I see **sustainability as a journey**. I believe that what customers all over the world expect from responsible companies is to know that they are moving in the right direction, keeping an eye on sustainability, and making their product's footprint smaller and smaller.

This book has three main sections:

Section I: The Case for Greener Products
Section II: Making Greener Products
Section III: Marketing Greener Products

Why is it an imperative to bring greener products to market? This question is answered in Chapters 1 through 3. We get a deeper understanding of the ecosystem pressures on the Earth and explore the market pressures being placed on businesses. This includes customers who buy products in supermarkets, business customers, and government purchasers who created a big demand of manufactures to bring more sustainable brands to market. People throughout the world are more health conscious and hear reports about the effects of chemicals on their health; therefore, they are seeking more organic and sustainably sourced products.

Global companies such as Walmart, Tesco, Lowe's, and IKEA, among others, have embraced sustainability and are triggering other companies to build eco-innovation into their product offerings as well. Governments have responded by putting regulatory pressure on product developers to initiate tougher new requirements such as removing materials of concern (e.g., PVC, triclosan, brominated flame retardants, BPA), reduce the carbon footprint, design more recyclable packaging, and facilitate product take-back at its end of life. These global regulations started out in Europe and have exploded throughout the world, making it very complicated for businesses to keep up with the ever-changing requirements and be compliant.

How do you make greener products? To find out how to go about making a greener product, it's a good idea to evaluate the best practices from leading companies. Chapter 4 covers best practices in a variety of industrial sectors, such as apparel, chemicals, electronics, consumer packaged goods, medical products, and energy.

I am privileged to have one of the leading thinkers and practitioners in developing greener products, Jim Fava, who will bring his wisdom from decades of product improvement work. Jim shares his thoughts on frameworks used to make products more sustainable and the golden rules for sustainable innovation in Chapter 5. Leading practices are reviewed, and lessons learned on how to address the myriad of demands on companies noted above are discussed.

Libby Bernick discusses the importance of considering biodiversity and natural capital in Chapter 6. Libby explains why we need to think about the cost to nature of the products we manufacture. Examples of companies that have used natural capital to make decisions on improving their product's sustainability performance are illustrated in the chapter.

A greener product is useless if no one knows it exists! **Appropriate marketing** of eco-improved products is a critical aspect of a sustainability program. Why should we consider green marketing? It's the secret sauce that brings greener products to life. The third section of this book addresses *green marketing*—our evaluation of sustainable brand marketing begins by reviewing some exemplary case studies in Chapter 7. We look at some of the best examples of green marketing from companies that have been very successful in making inroads with their customers.

Data on consumer behavior indicate that all over the world, there is a desire to buy products that are made by responsible companies and have greener attributes. But most do not want to pay more for them. If companies want to be successful in selling greener products, they cannot be at a higher price point, and they should sell them with an "**and**" in mind—an effective product at the right price *and* **it's greener**.

Not only must consumer marketers be concerned about bringing eco-innovative products to market but also business-to-business (B2B) marketers must be savvier in greener product offerings. Scorecards by companies like Walmart, Kaiser Permanente, and Procter & Gamble and the advent of business to business (B2B) green purchasing make the case that all product marketers must pay attention to making and marketing greener products.

Based on my analysis of successful green marketing campaigns, there are **three keys to winning**:

1. Have a credible greener product story.
2. Meet your customers' greener product demands.
3. Appropriately communicate the products' greener attributes.

These three characteristics will be used to evaluate the approach of several leading companies in business-to-consumer (B2C) and B2B marketing. Some of their initiatives have been groundbreaking and have paved the way to the current state of green marketing. We evaluate what made Ecomagination so impactful, how Clorox Green Works changed the game for mainstreaming green marketing, as well as innovative approaches by companies like Honest Tea and Patagonia.

Suzanne Shelton who is a leading expert in green marketing discusses how consumer interest in more sustainable products continues to grow in Chapter 8 and shares market research that her firm developed. The desire for greener product attributes differs from global region and by generation and understanding the messages that resonate with your target market is critical. Putting customers into marketing focus group segmentation is helpful to position your product the best possible way for success.

Understanding a good framework for sustainable brand marketing from a leading marketing and communications firm, OgilvyEarth, will be helpful to anyone interested in marketing green. That is exactly what you will find in Chapter 9. There are specific steps that can be taken to enable the marketer to have a successful campaign and examples and case studies of effective company programs.

There are certain elements that can make a marketing campaign more effective and certain things that will blow it up. That's what Chapter 10 explores; helpful elements like the use of eco-labels are reviewed, along with understanding regulatory requirements. There are key aspects that must be understood such as the Federal Trade Commission (FTC) green guides and the UK DEFRA green marketing rules. Perhaps, even more important than the regulatory requirements are the avoidance of greenwashing and a newer concern, green hushing. Inappropriate use of marketing claims can be a death blow to a brand—it's critical to avoid greenwashing; evaluating The Seven Sins of Greenwashing will help. Several methods for marketing the right way, to enhance and protect your brand, are discussed.

Finally, Chapter 11 concludes with an evaluation of the best practices that leading companies have for making and marketing greener products. Through an evaluation of the various company initiatives discussed in the Making of Greener Products section of the book, we see that there are some commonalities among the companies that are best at developing eco-improved products. Use of all or a combination of these activities will enable any company to bring more sustainable products to the market.

A review of the common approaches of successful green marketing campaigns makes it clear to marketers what elements should be considered to have a successful approach to reaching customers. After all, what good is making improvements to a product if you don't effectively communicate it to your customers?

Making and marketing greener products is no longer a nice *to do*—it is a business imperative. I believe this book makes it clear that **the world needs greener products** and it's up to the marketers, R&D, and product stewardship leaders to make that happen. No matter what industry you're in, it makes business sense to institutionalize these concepts. Evaluating the leading practice examples and techniques in this book will give any product developer or marketer some ideas that they then can translate into their own company culture.

I am hopeful that this book will not only benefit business but will also enable students to better understand what it takes to make and market a greener product and will also be useful for academics and NGOs. We all are concerned about becoming more sustainable, and I believe that it is possible to strike a balance in meeting the world's product needs while appropriately reducing unsustainable practices to meet this demand.

DISCLAIMER

The views and opinions expressed in this book are those of the author and do not necessarily reflect the official policy or position of the Estée Lauder Companies.

Al Iannuzzi (Ph.D.)

About the Author

 **Al Iannuzzi**, Ph.D., is Vice President of Sustainability at The Estée Lauder Companies where he leads enterprise-wide efforts and establishes and executes sustainability goals and objectives. He also oversees environmental sustainability programs, human rights, ESG ratings, product sustainability, and climate and energy initiatives. He is also an Adjunct Professor at Indiana University & Purdue University, where he teaches Product Improvement and Sustainability. Al has more than 35 years of experience in the Environment, Health, Safety, and Sustainability field. Before working at The Estée Lauder Companies, he worked for Johnson & Johnson, where he led Design for the Environment and Green Marketing programs and developed sustainability strategies for the Consumer Product, Medical Device, and pharmaceutical sectors. He served as the chief architect of the Earthwards® greener product development program. Al has also worked as an environmental consultant and as a regulator for the New Jersey Department of Environmental Protection. Al has authored four books and has written numerous articles on sustainability and product stewardship.

Acknowledgments

I am truly privileged to have the opportunity to write the third edition for Greener Products. My purpose for writing this book was to encourage companies to make the world a better place and to demonstrate how others have done it through making and marketing truly greener/more sustainable products. Writing a book takes a lot of effort, and it's particularly challenging when you're working full time, but I am extremely thankful for having an exceptional team that helped and contributed to what I hope will be a book that makes a difference.

First, I must thank my family whom I love immensely. Before I start anything that will be time-consuming, I request permission from the most important person in my life, my wife Ronnie, who doubled as an advisor on this project, so thank you, Ronnie, for the approval, understanding, and help. Next, I would like to thank my daughter-in-law Danielle, who helped with research and writing for this book, especially Chapter 8—I appreciate you and glad that we finally have someone else in our family with a degree in environment and sustainability!

Having a team of super interns from Fordham University was a huge blessing, so I thank you Ora Kalaja for the great work on Chapters 2, 3, and 4, Alison Wang thank you for updating Chapters 9, 10, and 11 and Kim Louise Hayuga for helping on Chapter 6 and on the marketing plans for the book. All of you have a bright future ahead of you, you are the ultimate "go getters" and now you know a ton more about sustainability!

A great boost to Greener Products is the fantastic thought leaders who contributed to this effort. Jim Fava, a long-standing friend, and mentor who has contributed to all three editions of Greener Products. Your work on LCA, product sustainability, and the Golden Rules for sustainable product design is fabulous in Chapter 5. Jim is a trusted advisor and a catalyst in the world, encouraging companies to be more sustainable.

Libby Bernick did a wonderful job with Chapter 6 on Natural Capital, bringing in a new focus on biodiversity and demonstrated her leadership in this quickly developing area.

Having someone with amazing sustainable product marketing insights as Susan Shelton, the CEO of Shelton Group, adds tremendous knowledge to the marketing section of the book. Thank you for the wisdom you presented in Chapter 8.

Getting some great marketing know-how from the work of expert communicators/marketers from Ogilvy was an added benefit to Greener Products. John Jowers and Ivellisse Morales provided excellent clarity on how to effectively position your product using proven techniques from their experience at OgilvyEarth in Chapter 9.

Finally, I would like to thank Taylor & Francis for supporting me in a third edition of *Greener Products*, which will encourage others to pursue making their brand more sustainable.

We all are stewards of the earth—let's make the world better with Greener Products. (Genesis 1:26).

Then God said, "Let us make man in our image, after our likeness. And let them have dominion over the fish of the sea and over the birds of the heavens and over the livestock and over all the earth and over every creeping thing that creeps on the earth" (Genesis 1:26).

Moreover, it is required in stewards that one be found faithful (1 Corinthians 4:2).

Al Iannuzzi is Vice President, Sustainability at The Estée Lauder Companies, where he leads enterprise-wide efforts to establish and execute on sustainability goals and objectives. He leads the companies Planet and Product sustainability initiatives which include addressing ESG ratings and ratings, climate and energy, Social Impact & Sustainability reporting, Human Rights, Green Chemistry, Sustainable Packaging, and Responsible Sourcing. He is also an Adjunct Professor at Indiana University/Purdue University, where he teaches Product Improvement and Sustainability.

Al has more than 35 years of experience in the Environment, Health, Safety, and Sustainability field. Prior to working at The Estée Lauder Companies, he worked for Johnson & Johnson, where he led Design for the Environment and Green Marketing programs and developed sustainability strategies for Consumer Product, Medical Device, and Pharmaceutical sectors. He served as the chief architect of the Earthwards® greener product development program. He has also worked as an environmental consultant and as a regulator for the New Jersey Department of Environmental Protection.

Section I

The Case for Greener Products

1 Introduction

Al Iannuzzi

This introductory chapter provides an overview of the case for companies to consider sustainability and the driving forces behind the need to make and market greener products or more sustainable products. The purpose of this chapter is to cover some of the current events that are encouraging companies to pursue greener products. The rise of the global middle class is occurring at a rapid clip, with an additional 3 billion people expected to enter the consumer market by 2030. The growing urban middle class brings with it an unprecedented demand for resources, putting more strain on the global supply of raw materials, fossil fuels, food, and water. Other drivers discussed are the UN Sustainable Development Goals (SDGs), environmental damage created by companies, and the mainstreaming of greener products. Some company examples are discussed from Tesla, IKEA, Coca-Cola, and PepsiCo.

SUSTAINABILITY BECOMES A BUSINESS IMPERATIVE

It has been a steady climb, and it is obvious that sustainability has become a mainstream business imperative. You would be hard pressed to find a multinational company that does not publish an annual sustainability report which has significant commitments in the areas of environment and social issues; and this is completely voluntary! There has been a rapid increase in Environment, Social and Governance (ESG) stock funds which have a sustainability bar that companies must meet to have their stock purchased for these funds. There are numerous ratings of companies ESG performance, such as the Dow Jones Sustainability Index, which has been closely connected to corporate reputation. Companies have responded to these trends, and we have seen sustainability as a frequent topic during Board meetings and a rapid growth of Chief Sustainability Officers (CSO) within companies, many of which report directly into the CEO. A survey of 500 C Suite executives concluded that all respondents believe sustainability and ESG issues are important to their organization, "with 87% believing those initiatives are very to extremely important to their businesses and long-term success" (EY 2023).

In addition to these trends, we see the growth of the global middle class is occurring at a rapid clip, with an additional 3 billion people expected to enter the consumer market by 2030 (Nyquist et al. 2016). The growing urban middle class brings with it an unprecedented demand for resources, putting more strain on the global supply of raw materials, fossil fuels, food, and water. Countries like India and many in Africa are wanting to electrify and increase their energy demand significantly. Furthermore, greenhouse gas emissions and waste generation will continue to soar as global consumption increases. With this new reality apparent, **"business as usual" is no longer an option**. Resources will have to be used in a more efficient

DOI: 10.1201/9781003441939-2

way to meet the needs of global population, both present and future and this affects product development.

Under the pressure of these growing demands, the quest for sustainability has transformed the business landscape. Sustainability that was once viewed as a simple risk-management function is now regarded as a business opportunity. The growing demand for resources offers a chance for business leaders to meet demand in a more sustainable way and in turn revolutionizes how we think about consumption. Businesses are even shifting their missions and values to align with their sustainability goals, with 9 out of 10 executives reporting board oversight of their organizations' sustainability and ESG agendas (EY 2023). Greener products are playing a bigger role in business than ever before. Sustainability is no longer some idealistic notion; it is at the forefront of development and is even being used as an innovation driver. Company leaders are rallying behind sustainability, and **things will never be the same**.

C- SUITE INSIGHTS

- 87% believe sustainability and ESG initiatives are very to extremely important.
- 82% their organizations have both carbon emissions reductions initiatives in place and goals to reach net zero by a given year.
- 81% have a CSO or equivalent position.
- 90% have board oversight of their organizations' sustainability and ESG agenda.

EY, 2023

C-suite Insights: Sustainability and ESG Trends Index,
Survey of 500 C – Suite Executives

Based on my experience at the Estée Lauder Companies we focus on the key stakeholder groups of our consumers, investors, and employees. All of these groups have increasing interest in sustainability. When I started in 2018 there were just a few calls with investors on sustainability, and in 2023 there were 30+ meetings, investors are constantly asking about how we manage issues like biodiversity, deforestation, social issues in the supply chain, plastic usage, and of course our climate programs. Our consumers are concerned with the ingredients we use in our cosmetic products and if the packaging can be recycled at the end of its life. Our employees want to work for a company that is a good corporate citizen and is addressing some of the biggest issues that are important to them like climate change and social issues such as gender pay equity and hiring practices that are inclusive of all.

WHAT CAUSED THIS SHIFT?

ENVIRONMENTAL HISTORY

Environmentalism has appeared sporadically throughout history, in many different forms, all over the world. Many associate the beginning of the modern environmental movement with Earth Day and the legislative fervor of the 1970s, although its origins are rooted a bit more deeply in the past. "In wildness is the preservation of the world," wrote Henry David Thoreau in his nature-themed essay *Walking*. Thoreau and the transcendentalists of the mid-1800s were some of the first to suggest preserving the natural world for its beauty and potential for spiritual enlightenment, not to mention its immense practical value. These philosophers realized that humans rely heavily on nature for their survival, and the nation's natural bounty was not infinite. The 1860s brought the naturalist John Muir and the creation of the first national park in the United States. The National Park Service was created soon after, in 1916 – recently celebrating its 100th anniversary. Although "going green" may seem like a recent trend, environmental consciousness is deeply woven into the fabric of world history.

In the wake of World War II, environmentalism quickly took a backseat to growth and innovation. The postwar economic boom brought with it increased consumption, waste, and pollution. The pollution continued largely unchecked until the 1960s brought a new wave of environmentally conscious thinking, catapulted by Rachel Carson's landmark book *Silent Spring*, which helped set the stage for the environmental movement by exposing the environmental hazards of pesticides and by questioning humanity's unwavering faith in technological advancement. In June 1972, the United Nations Conference on the Human Environment was held in Stockholm, Sweden. This conference would lay the foundation for modern global environmental policy. The 1970s also brought a rapid series of new environmental legislation, the creation of the US Environmental Protection Agency, and the inaugural celebration of Earth Day. Thus began the environmental movement of today. It rapidly gained momentum over the following decades and exploded onto the world business scene and has taken on the term "**sustainability**." The focus on more sustainable products is due, in part, to greater awareness of major environmental issues such as climate change and plastic pollution and consumers wanting to make a difference with their purchasing decisions.

Global Environmental Risks Over the Next 10 Years

- Failure to mitigate climate change.
- Failure of climate change adaption.
- Natural disasters and extreme weather events.
- Biodiversity loss and ecosystem collapse.
- Natural resources crisis.
- Large scale environmental damage incidents.

(WEC 2023)

Although tremendous gains have already been made to address environmental issues there is a lot more work to be done and consumers and businesses are more aware than ever. Some disturbing trends that are fueling this awareness and making it more important to take action are: by 2030 renewables will remain a mere fraction of our energy supplies, 660 million people will remain without electricity and close to 2 billion will continue to rely on polluting fuels and technologies for cooking. A great number of species worldwide are threatened with extinction and 670 million

FIGURE 1.1   Sustainable Development Goals. (From United Nations 2024.)

people will still be facing hunger – 8% of the world's population. In 2022 there are 2.2 billion people still lacking safely managed drinking water services, 3.4 billion lack safely managed sanitation services, and 1.9 billion lack basic hygiene services. And the ocean continues to be endangered by rising acidification, eutrophication, declining fish stocks, and mounting plastic pollution (UN Progress towards the Sustainable Development Goals p. 2, 3, 8, 12, 20, 2023).

To tackle these global issues, world leaders assembled at the United Nations Sustainable Development Summit in September 2015 to adopt the 2030 Agenda for Sustainable Development, which includes a set of 17 SDGs (Figure 1.1), effectively replacing the Millennium Development Goals that were in place from 2000 to 2015. These goals are targeting the most pressing environmental and humanitarian issues in the world today.

UN SUSTAINABLE DEVELOPMENT GOALS

1. End poverty in all its forms everywhere.
2. End hunger, achieve food security and improved nutrition, and promote sustainable agriculture.
3. Ensure healthy lives and promote well-being for all at all ages.
4. Ensure inclusive and equitable quality education and promote life-long learning opportunities for all.
5. Achieve gender equality and empower all women and girls.
6. Ensure availability and sustainable management of all water and sanitation for all.
7. Ensure access to affordable, reliable, sustainable, and modern energy for all.
8. Promote sustained, inclusive, and sustainable economic growth; full and productive employment; and decent work for all.

9. Build resilient infrastructure, promote inclusive and sustainable industrialization, and foster innovation.
10. Reduce inequality within and among countries.
11. Make cities and human settlements inclusive, safe, resilient, and sustainable.
12. Ensure sustainable consumption and production patterns.
13. Take urgent action to combat climate change and its impacts.
14. Conserve and sustainably use the oceans, seas, and marine resources for sustainable development.
15. Protect, restore, and promote sustainable use of terrestrial ecosystems, sustainably manage forests, combat desertification, halt and reverse land degradation, and halt biodiversity loss.
16. Promote peaceful and inclusive societies for sustainable development, provide access to justice for all and build effective, accountable, and inclusive institutions at all levels.
17. Strengthen the means of implementation and revitalize the global partnership for sustainable development.

Sustainable Development Goals. (From United Nations 2024.)

The SDGs have stirred up significant activity; government leaders worldwide have seemingly embraced the new goals, with numerous presidential decrees, national action plans, new policies, budgets, and stakeholder collaboration platforms being rolled out in direct response to the SDGs. Businesses have also taken notice, and many are incorporating the SDGs into their objectives with surprising seriousness. The United Nations (UN) has developed an objective to encourage companies to participate in the SDGs called the "Global Compact." This initiative supports companies and requests that they commit to 10 principals and take action on the SDGs (UN Global Compact 2023). It is fairly common for companies to mention how they are supporting the SDGs in their annual sustainability report. The company I work for, Estée Lauder companies, is a prestige beauty company that makes cosmetics and we speak to the actions we take to support the SDGs. If you think about the operations of a cosmetic company that was founded by a woman in the 1940s and employs 80+% woman, then it makes sense to support: SDG 3 – Good Health and Well Being, SDG 5 – Gender Equality, SDG – 10 Reduce Inequalities, SDG – 12 Responsible Consumption & Production, SDG 13 – Climate Action, and SDG – 15 Life on the Land (ELC 2022 p. 20). It doesn't make sense to me to set objectives for all 15 SDGs, but rather to focus on the ones that are most closely connected to your business.

It is common for companies to mention how they are supporting the SDGs in their annual sustainability report.

CONSTANT PRESSURES ON THE ENVIRONMENT

A constant barrage of environmental catastrophes are being witnessed by the world, and the stories never seem to stop coming. "As of May 2023, CO2 PPM (parts per million) is at 420.00 and the global temperature rise is reported to be 1.15C compared to pre-industrial levels." This increase in CO_2 is claimed to be driving many environmental issues like forest fires, ice cap melting, and severe storms. Food waste is a significant issue, especially when you consider the resources to grow the food like the use of fertilizer, herbicides, and water. It's amazing to think that a "**third of the food intended for human consumption – around 1.3 billion tons – is wasted or lost**. This is enough to feed 3 billion people." And when you consider emissions from this waste, it's a full third of all greenhouse gas emissions annually! A World Wildlife Fund report indicated that the population sizes of mammals, fish, birds, reptiles, and amphibians have experienced a decline of an average of 68% between 1970 and 2016. This biodiversity loss is attributed to land-use change, particularly the conversion of habitats, like forests, grasslands, and mangroves, into agricultural systems. In addition, more than 500 species of land animals are on the brink of extinction. In 1950, the world produced more than 2 million tons of plastic per year, and this increased dramatically to 419 million tons which has contributed to the daunting issue of plastic waste in the environment, especially that which makes its way into the ocean (about 14 million tons/year). Plastic makes its way throughout the ocean, including being ingested by sea life. **If no action is taken, ocean plastic will grow to 29 million metric tons per year by 2040!** If we include microplastics into this, the cumulative amount of plastic in the ocean could reach 600 million tons by 2040, this is to me the second largest environmental issue next to climate change. All the plastic issues are connected to this statistic, **roughly 91% of all plastic made is not recycled.**

> In 1950, the world produced more than 2 million tons of plastic per year, and this increased dramatically to 419 million tons which has contributed to the daunting issue of plastic waste in the environment, especially that which makes its way into the ocean – about 14 million tons/year.

Deforestation is associated with many environmental issues, the loss of biodiversity, and carbon sequestering for example. The countries experiencing the highest levels of deforestation are Brazil, the Democratic Republic of Congo, and Indonesia. "The Amazon, the world's largest rainforest – spanning 6.9 million square kilometers (2.72 million square miles) and covering around 40% of the South American continent – is also one of the most biologically diverse ecosystems and is home to about **three million species of plants and animals**." Despite efforts to protect forest land, legal deforestation is still rampant, and about a third of global tropical deforestation occurs in Brazil's Amazon Forest, amounting to **1.5 million hectares each year.** Air pollution is still a very significant issue of concern, especially in developing nations. The World Health Organization estimates that **4.2 to 7 million people die from air pollution worldwide every year.** "Causes of air pollution mostly comes from industrial sources and motor vehicles, as well as emissions from burning biomass and poor air quality due to dust storms." Finally, ocean acidification can result in "irreversible changes in habitat quality." If pH levels go too low, marine organisms such as oysters can be severely impacted. But one of the acidification issues that are of most concern is the bleaching of coral reefs, a very significant support of marine species. Also, **oceans absorb about 30% of carbon dioxide** that is released into the Earth's atmosphere. As more carbon emissions are released into the environment, the ocean is going to have to work harder to absorb these emissions (Earth.Org 2023).

There is little doubt that these disasters shape our thinking about the environment and the pressure we are putting on it. Consumers show their anger about pollution by punishing corporate violators by not purchasing their products or buying their stock. Global environmental damage caused by human activity was estimated at $6.6 trillion and about 11% of global gross domestic product in 2008. It is estimated that the **top 3,000 public companies are responsible for one third of global environmental damage**; it is no wonder citizens everywhere want to move their buying power toward environmentally conscious companies and away from those perceived as irresponsible (Environmental Leader 2010). The desire of consumers to "make a difference" with their purchases is growing, and it is imperative that companies are aware of current environmental and resource issues and minimize the impact of their products. Companies also need to begin looking beyond the surface, as the impact of products stretches deep into the supply chain. This includes raw materials that are harvested or taken from the earth.

Companies are giving all these issues serious consideration when conducting business. The consulting firm Deloitte wrote the report, The Fourth Industrial Revolution: At the Intersection of Readiness and Responsibility. They clearly made the point that environmental impact and sustainability issues have become critical to business operations. Surveying more than 2,000 global executives, Deloitte Global found that almost 90 percent agreed to some degree that the impacts of climate change will negatively affect their organizations. Nearly six in 10 claimed to have internal sustainability initiatives in place, from reducing travel to eliminating plastics, and more. Executives are taking notice of these issues and consider sustainability programs as a business imperative as they feel more pressure from consumers and other stakeholders (Deloitte Insights 2020).

> Executives are taking notice of these issues and consider sustainability programs as a business imperative as they feel more pressure from consumers and other stakeholders.

MILLENNIALS AND GEN Z DEMAND ACTION

There has been a growing interest in seeing companies do more than just make a profit, but rather to operate in a responsible manner, as a good corporate citizen. This sentiment is especially true with younger consumers; "Mllennials and Gen Zs are looking for more than transactional relationships with companies. Whether they're buying candy bars or jeans, consumers increasingly demand that businesses do their part to reduce their environmental impact more than one in four millennials and Gen Zs believe businesses should try to help mitigate the effects of human-caused climate change and protect and improve the environment. Yet only 12 percent believe corporations are working to address things like climate change."

Large numbers of young consumers have started or stopped relationships with businesses based on their perceptions of companies' commitments to society and the planet. "**Almost 40 percent of those asked said they would stop buying from a company whose products or services negatively impact the environment**" (Forbes 2023). Companies that do not address these trends will be losing an important purchasing group today and in the future.

BUSINESS IS RESPONDING

It is evident that there is a growing desire for greener products, a Deloitte survey indicated that nearly two-thirds of companies said their customers have been demanding they switch to renewable sources of electricity (Forbes 2023). Evidence of the increased attention being given to sustainability is the voluntary reporting on such issues as greenhouse gas emissions, deforestation prevention, and water use. The **CDP**, a nonprofit organization that maintains a database where **companies voluntarily report environmental progress**, is a very relevant example. Data is provided to investors and firms that rate the sustainability performance of companies. Nearly 15,000 companies submitted data on climate change, forests, and water security progress in 2022. Companies complete very detailed questionnaires and provide data on such things as tons of GHG emissions and are rated on an A to D scale. Getting an "A" on one of these surveys is not an easy feat, and it is considered a proof point that you are making serious progress to manage your environmental impact (CDP 2023). And yes, all this is done voluntarily; market pressures are compelling this reporting.

Perhaps an even greater demonstration of corporate commitment to addressing environmental issues are the 400+ companies that have made the very daunting commitment to source 100% of their electricity from renewable sources. The NGO the Climate Group runs the program which is called **RE100** along with an accompanying initiative for a commitment to 100% use of electric vehicles (EVs) **EV100** (TCG 2023). I can tell you that being able to make this claim is very meaningful with our customers and demonstrates that our company is serious about addressing some of the most pressing environmental concerns out there.

ENVIRONMENTAL PRACTICES OF SHOPPERS

Sixty percent of global consumers say they are already following these sustainability practices.

* Recycling products, bottles and packaging.
* Using reusable cloths for cleaning.
* Buying refillable cleaning and home care products.

BCG, 2023

*Consumers are the Key to Taking Green Mainstream.
Countries surveyed include the United States, Brazil,
China, France, Germany, Italy, India, and Japan*

MAINSTREAMING OF GREENER PRODUCTS

Pressure on businesses to "go green" is growing rapidly, but it is not coming from the government. Perhaps surprisingly, the largest driving force behind more sustainable product design is in fact what I like to call **market drivers.** A prime example is the world's largest retailer **Walmart**. Once Walmart started asking for more sustainable merchandise, it began a boon for greener products. Being such a large retailer, its demands for greener products were momentous and far-reaching and the effects stretched deep into the supply chain as well as catalyzed the competition.

Once Walmart started their sustainability journey, many other companies followed suit by setting their own sustainability goals. Competition ensued between rival companies, further pushing the boundaries of sustainable design. Due to the emerging demand, sustainability had become a key driver of innovation. More sustainable products have cropped up in all aspects of business, trying to address the consumer's desire for more ethical goods. In turn, they provide an additional benefit, such as providing safer products for the home or saving money through energy efficiency.

Once Walmart started asking for more sustainable merchandise, it began a boon for greener products. Being such a large retailer, its demands for greener products were momentous and far-reaching.

Sustainability market pressures have been affecting all industrial sectors. An early example of this was how the automotive industry had been revolutionized with the success of the **Toyota Prius** hybrid. Being the first car that had a green platform from its inception, it has been well received by a public that is eager to purchase a high fuel-efficiency product. Though its success was likely due in part to the rapidly rising gas prices of the early 2000s, the sale of electric and hybrid vehicles is still on the rise in both developing and developed countries (Kaye 2016).

Further highlighting this market-based revolution, the **Tesla Model 3** garnered a lot of attention when it was revealed in April 2016. The so-called "electric vehicle for the masses" hits the sweet spot-on price, design, and performance – and people were excited to buy an EV. Over 180,000 pre-orders were placed in the first 24 hours, with each reservation requiring a hefty $1000 deposit. Where traditional automakers typically design eccentric, unconventional EVs marketed toward environmentalists (think Nissan Leaf), Tesla's "think-big" attitude may have resulted in the first true "everyman's" electric car. It amazes me how Tesla upended the entire automobile market. Prior to the advent of Tesla, would you have thought that a new automotive company would have been successful, and not only that but that they would revolutionize the entire industry based on a sustainability platform? This is a market driver at its best – no government regulation brought Tesla into existence, it was just the ambition of an entrepreneur. Now, every major global car company has committed to developing EVs. Sales for EVs are growing at a rapid pace; global sales increased 55% in 2022 compared to the previous year reaching a total of 10.5 million. These figures include both battery EVs and plug-in hybrids (WEC 2023). I have seen estimates that by 2030 around 50% of all vehicle sales will be EVs too, including the one I work for. We have signed on to The Climate Groups EV100 initiative where we committed to convert our entire fleet to EVs by 2030.

Beverage behemoths **Coca-Cola** and **PepsiCo** have also jumped on the sustainability bandwagon. Water is the primary ingredient in their beverages, and both companies recognize that the increasing stress on water supply is a risk to their business. Coca-Cola has made significant commitments to address this very important resource for their business. Achieve 100% circular water use – or regenerative water use – across 175 facilities identified as "leadership locations" by 2030. Work with partners to help improve the health of 60 watersheds identified as most critical for the company's operations and agricultural supply chains by 2030. They **plan to return a cumulative total of 2 trillion liters of water to nature and communities globally**, between 2021 and 2030. Meanwhile, PepsiCo has committed to address high water-risk areas, by 2025 to: improve water use efficiency by 15% in their agricultural supply chain (focused on corn and potatoes), improve operational water use efficiency by 25%, replenish 100% of the water used in manufacturing operations, follow the Alliance for Water Stewardship standard, and deliver safe water access to 25 million people by 2025 (Pepsico 2023).

The largest furniture retailer in the world, **IKEA**, has significant sustainability commitments which they call **People & Planet Positive** with three key focus areas:

1. *Sustainable living*: Inspiring and enabling more than 1 billion people to live better lives within the limits of the planet.
2. *Circular & Climate Positive*: Becoming climate positive and regenerating resources while growing the IKEA business.
3. *Fair & Equal*: Creating a positive social impact for everyone across the IKEA value chain.

The results so far are promising since 2015: 100% of the cotton, fish, seafood, wood & paper sold come from more sustainable sources.. And they have taken steps to begin to phase out virgin fossil plastic from the IKEA product range, a focus that will also continue toward 2030. I really like their commitments to the circular economy. IKEA has committed to prolong the life of products and materials, using resources in a smarter way. Turning waste into resources, sending zero waste to landfill and taking the transforming secondary materials (i.e., reused and recycled materials) into clean and safe resources. Establishing and promoting systems and services to enable a circular economy (Ikea 2023).

GREEN BUILDINGS

Another major example of the mainstreaming of greener products is the green building movement. It is commonplace to see Leadership in Energy and Environmental Design (LEED) plaques when you enter an office or manufacturing building. Businesses have come to realize that there is a good reason to pursue green building concepts for new and renovated buildings. The **buildings and buildings construction sectors are "responsible for over one-third of global energy consumption**

and nearly 40% of total direct and indirect CO2 emissions." Green buildings help improve indoor air quality, reduce exposure to toxins and pollutants, and provide access to outdoor spaces and daylight, which has important mental and physical benefits. Operating cost savings, shorter payback periods, and increased asset value have been consistently reported by those implementing green buildings. An estimate of **LEED buildings from 2015 to 2018 showed $715.3 million in maintenance savings** (USGBC 2023).

There has also been a steady increase in the use of solar on homes. Nearly 4% of all homes in the United States had solar photovoltaic (PV) in 2020. Businesses have also responded to this movement too and with about 1.6% of commercial buildings having small-scale solar generation. Use of PV solar is a great opportunity for companies to generate their own green power, and it resonates really well with employees – an obvious commitment to your sustainability goals. Following the implementation of the 2023 Inflation Reduction Act in the United States, a study by Princeton University indicated solar development could grow from 2020 rates of 10 GW of capacity per year to as much as **five times more** by 2024, adding 49 GW of utility-scale solar each year. Solar deployment may be well over 100 GW per year by 2030, according to this study (PV Magazine 2023). Many companies have set targets for LEED or equivalent certifications for their new buildings and are deploying on site solar photovoltaics on their operations. This is an objective that we brought forth at Estée Lauder Companies because it makes a lot of sense. It enables employees to experience your environmental commitments, there is a return on investment due to generating your own energy and it helps with the reduction of greenhouse gases. The green building movement is on and growing, another good example of a non-regulatory **market driver**.

WHY THE FOCUS ON GREENER PRODUCTS?

Annie Leonard makes a very good case for focusing on how we create, use, and dispose of products in her short documentary, *The Story of Stuff*. Despite being over a decade old at the time of this writing, the video still discusses some of the most relevant issues facing our consumption culture today. After its debut in 2007, it quickly became an Internet sensation with over 40 million views worldwide and spurred an entire series of similar movies and projects promoting social and environmental change. If you have not seen it, I would encourage you to view it at storyofstuff.org. A compelling case is made that products need to be managed in a much more sustainable manner (Story of Stuff 2023).

Traditionally, environmental management practices only focused on managing risks and reducing the carbon footprint at the manufacturing and production level. With the emergence of life cycle thinking (also known as cradle to grave) and life-cycle assessment methodology, we are now looking at the environmental, social, and economic impacts of a product over its entire life cycle. In fact, some of the greatest environmental benefits occur through the selection of raw materials, packaging, distribution, use phase, and disposal of a product.

One of the greatest product examples of life-cycle thinking is cold water laundry detergent, developed by Proctor & Gamble. Life-cycle assessments revealed that during the product use phase the heating of water in the home consumed far more energy than any other life-cycle stage (e.g., raw material acquisition, manufacturing, transportation, end of life). In response, a detergent that functions in cold water was introduced into the market, which does not require heating of wash water – smart! (P&G 2024).

CIRCULAR ECONOMY

A concept that I really feel is valuable because it mimics nature is that of a circular economy, initiated by the Ellen MacArthur Foundation (EMF) (Figure 1.2). The circular economy takes the idea of life cycle thinking one step further; the current "take, make, dispose" economic model is dependent on an endless supply of cheap raw materials and energy, as well as an unlimited capacity to store waste. **A circular economy is "restorative and regenerative" by design,** seeking to produce zero waste through an enhanced flow of goods and services. The way EMF defines it is: "the circular economy is a system where materials never become waste and nature is regenerated. In a circular economy, products and materials are kept in circulation through processes like maintenance, reuse, refurbishment, remanufacture, recycling, and composting. The circular economy tackles climate change and other global challenges, like biodiversity loss, waste, and pollution, by decoupling economic activity from the consumption of finite resources" (Ellen MacArthur Foundation 2023). This

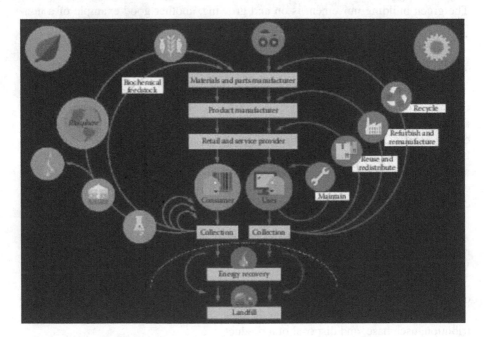

FIGURE 1.2 Ellen MacArthur circular economy. (With permission from Ellen Macarthur Foundation, 2023.)

resonates with anyone, like me, who is interested in protecting the environment and following the way nature operates.

The circular economy concept has exploded in popularity, with many companies embracing the idea. Among several other social and environmental initiatives, clothing brand Eileen Fisher, Inc. incorporated a circular economy into their business model through their line of recycled clothing "Green Eileen." The company began as a nonprofit in 2009, Green Eileen recycles clothing by reselling items in their Green Eileen stores. With the average American throwing out 70 pounds of clothing each year, Green Eileen extends the life of used garments by creating a second market, customers return used Eileen Fisher garments in any condition and receive a five-dollar gift card. Anything that cannot be resold is upcycled through a program they call "Remade in the USA," which turns flawed garments into entirely new designs. The program has been so successful that over 2 million garments have been taken back (Eileen Fisher 2023). Many other big-name brands, including Levi Strauss, Nike, Dell, Patagonia, IKEA, and H&M, are working to make the concept of a circular economy a reality, largely through innovations in the recycling of material.

> The most sustainable garment you can wear is one that's already been made, and already been worn.
>
> **Lilah Horwitz**
> *Director of Renew Content & Marketing, Green Eileen*

Adidas made a splash with an athletic shoe made almost completely of recycled plastic from oceans. With over 8 million tons of plastic entering the planet's oceans every year, Adidas partnered with Parley for the Oceans, the UltraBOOST Uncaged Parley, with an upper made from 95% ocean plastic collected from the Maldives. The shoelaces, heel cap base, heel webbing, heel lining, and the sock liner cover are also made with 100% recycled materials (Adidas 2016; McAlone 2016). Adidas planed for around 15 million sneakers in 2020. "This means less virgin plastic, reduced CO_2 emissions and more awareness of the issue – with every product made serving as a wearable symbol of change that sparks questions, discussions and ideas about creating even more progress." This is a great example of a product addressing an environmental problem, making a profit and doing good at the same time (Adidas X Parley 2023).

THINGS WILL NEVER BE THE SAME

Sustainability is here to stay, beginning as a voluntary business focus, we are seeing governments embedding its concepts into regulations (see Chapter 3, Regulatory Drivers for Greener Products). In December 2015, 195 countries at the 21st Conference of the Parties of the UNFCC adopted one of the most significant diplomatic agreements in history. The landmark Paris agreement is an **international treaty**, the first of its kind, to address climate change. On October 4, 2016, the European Union voted to ratify the Paris agreement, passing the 55-country threshold needed for the agreement to enter into force. A huge triumph for the planet, the Paris Agreement is a big step for worldwide sustainability, spurring the growth of renewable energy throughout the world.

Following the conference in Paris, France became the first country in the world to **ban plastic plates, cups, and utensils**. The new law, which went into effect in 2020, is a part of the country's Energy Transition for Green Growth Act. The same legislation banned the use of plastic bags in grocery stores and markets. According to French lawmakers, the goal is to promote a "circular economy" of waste disposal, "from product design to recycling" (McAuley 2016). In December 2015, President Barack Obama signed a bill prohibiting the sale of products containing microbeads. Often used in exfoliating and cleansing products, microbeads have been a huge concern for aquatic environments, where the beads eventually end up (Imam 2015). The United Kingdom quickly followed suit, pledging to ban them by 2017 (BBC 2016). The **UN held a Biodiversity Conference** (COP 15) in Montreal Canada on December 19, 2022, with an action plan for nature to 2030. The plan includes actions to halt and reverse nature loss, putting 30% of the planet and 30% of degraded ecosystems under protection by 2030 (UN Environment Program 2023).

Government regulation can progress sustainability however, consumers will continue to be the main driver for more sustainable products, and research shows they are demanding it. According to a study from the consulting group Deloitte, in the United Kingdom 65% are looking for products that are made from natural or renewable materials and 56% want responsibly sourced products (Deloitte 2023). Demand for environmentally friendly products is only expected to grow in the future as the younger generations increase their buying power. Globally, 85% of people indicate that they have shifted their purchase behavior toward being more sustainable in the past five years (Businesswire 2023). **Things will never be the same because greener products are not just a nice to do, they are a business imperative!**

REFERENCES

Adidas. 2016. *Parley.* http://www.adidas.com/us/parley (Accessed November 16, 2016).
Adidas X Parley. https://parley.tv/initiatives/adidasxparley (Accessed October 29, 2023).
BBC News. 2016. *Plastic Microbeads to Be Banned by 2017, UK Government Pledges.* September 3. http://www.bbc.com/news/uk-37263087 (Accessed September 5, 2016).
BCG. Consumers Are the Key to Taking Green Mainstream. https://www.bcg.com/publications/2022/consumers-are-the-key-to-taking-sustainable-products-mainstream (Accessed July 30, 2023).
Businesswire. 2023. Recent Study Reveals More Than a Third of Global Consumers Are Willing to Pay More for Sustainability as Demand Grows for Environmentally-Friendly. (Accessed October 30, 2023).
CDP. https://www.cdp.net/en/companies/companies-scores (Accessed July 17, 2023)
Coco-cola. https://www.coca-colacompany.com/media-center/2030-water-security-strategy (Accessed October 21, 2023).
Deloitte. Sustainable Consumer 2023. https://www2.deloitte.com/uk/en/pages/consumer-business/articles/sustainable-consumer-what-consumers-care-about.html (Accessed October 27, 2023).

Deloitte Insights. 2020. The Forth Industrial Revolution. 2020 Deloitte Development LLC.

Earth.Org. 2023. 15 Biggest Environmental Problems of 2023. https://earth.org/the-biggest-environmental-problems-of-our-lifetime/ (Accessed July 17, 2023).

Eillen Fisher. https://www.eileenfisher.com/a-sustainable-life/journal/sustainability/renew-program-reaches-2-million-garments.html (Accessed May 12, 2024).

Ellen MacArthur Foundation. 2023. What is a *Circular Economy?* https://www.ellenmacarthurfoundation.org/topics/circular-economy-introduction/overview (Accessed October 28, 2023).

Environmental Leader. 2010. *Top Public Companies Cause One-Third of the World's Environmental Damage.* October 5. http://www.environmentalleader.com/2010/10/05/top-public-companies-cause-one-third-of-the-worlds-environmental-damage/#ixzz4PS2BPHm1 (Accessed October 1, 2016).

Estée Lauder Companies (ELC) Fiscal 2022 Social Impact and Sustainability Report, p. 20.

EY. 2023. C-suite Insights: Sustainability and ESG Trends Index.

Forbes. Reducing Environmental Impact Is Now a Business Imperative. https://www.forbes.com/sites/deloitte/2020/01/22/reducing-environmental-impact-is-now-a-business-imperative/?sh=491334fb6cc6 (Accessed November 4, 2023).

Ikea. https://www.ikea.com/us/en/this-is-ikea/about-us/the-ikea-sustainability-strategy-making-a-real-difference-pubb5534570 (Accessed October 21, 2023).

Imam, Jareen. 2015. Microbead ban signed by President Obama. *CNN.* December 31. http://www.cnn.com/2015/12/30/health/obama-bans-microbeads/ (Accessed September 5, 2016).

Kaye, Leon. 2016. Despite Low Oil Prices, Electric Vehicle Sales Keep Rising. *Triple Pundit.* August 24. http://www.triplepundit.com/2016/08/despite-low-oil-prices-electric-vehicle-sales-keep (Accessed September 1, 2016).

McAlone, Nathan. 2016. Adidas Is Selling Only 7,000 of These Gorgeous Shoes Made from Ocean Waste. *Business Insider.* November 4. http://www.businessinsider.com/adidas-shoe-from-plastic-ocean-waste-2016-11/#-1 (Accessed November 16, 2016).

McAuley, James. 2016. France Becomes the First Country to Ban Plastic Plates and Cutlery. *The Washington Post.* September 19. https://www.washingtonpost.com/news/world-views/wp/2016/09/19/france-bans-plastic-plates-and-cutlery/ (Accessed October 2, 2016).

Nyquist, Scott, Matt Rogers, and Jonathan Woetzel. 2016. The Future Is Now: How to Win the Resource Revolution. *McKinsey Quarterly,* 4: 100–117.

Pepsico. https://contact.pepsico.com/pepsico/article/how-does-pepsico-conserve-and-protect-water (Accessed October 21, 2023).

P&G. 2024. Can washing your clothes on cold do a world of good? https://us.pg.com/blogs/pg-sustainability-tide-ariel-cold-water-wash/ (Available April 18, 2024)

PV Magazine. Nearly 4% of US homes have solar panels installed. https://pv-magazine-usa.com/2022/10/28/nearly-4-of-u-s-homes-have-solar-panels-installed/ (Accessed October 26, 2023).

Story of Stuff. 2016. *From a Movie to a Movement.* http://storyofstuff.org/ (Accessed October 29, 2023).

The Climate Group (TCG). RE100. https://www.there100.org/re100-members (Accessed July 17, 2023)

United Nations. 2023. Sustainable Development Goals. https://sdgs.un.org (Accessed July 22, 2023).

UN Sustainable Development Goals. https://sdgs.un.org/goals. Available April 17, 2024.

United Nations Environment Programme. COP 15 ends with landmark biodiversity agreement (Accessed October 27, 2023).

United Nations, Progress towards the Sustainable Development Goals: Towards a Rescue Plan for People and Planet (May 2023).

United Nations, Global Compact: Business as a Force for Good. https://unglobalcompact.org/
    what-is-gc/mission (Accessed July 19, 2023).
UN SDG Tracker: Measuring progress towards the Sustainable Development Goals https://
    ourworldindata.org/sdgs. Accessible April 19, 2024.
US Green Building Council (USGBC) Benefits of Green Buildings https://docs.google.
    com/document/d/1fPuNCmVmFlylDKP19_hI6DSWFzb6EHXOrALNZaRnYlI/edit
    (Accessed October 26, 2023).
World Economic Forum https://www3.weforum.org/docs/WEF_Global_Risks_Report_2023.
    pdf. (Accessed April 19, 2024).
World Economic Forum. Electric Vehicle Sales leapt 55%. https://www.weforum.org/
    agenda/2023/05/electric-vehicles-ev-sales-growth-2022/ (Accessed July 30, 2023).

2 Market Drivers for Greener Products

Al Iannuzzi

This chapter will discuss the market drivers that are requiring companies to develop more sustainable products. In 2022, Nielsen IQ revealed that "78% of US consumers say that a sustainable lifestyle is important to them." The study found that ESG claims on animal welfare, environmental sustainability, organic positioning, plant-based, social responsibility, and/or sustainable packaging, in general, play a significant role in consumer buying behavior leading to greater growth rates than products with no ESG claims (NielsenIQ 2022). Retailers continuously putting pressures on companies is one of the greatest drivers for developing greener products; when the marketplace demands it companies respond. This is especially true when it's your largest customer. Walmart and Sephora have developed significant sustainability requirements for their suppliers, as have home improvement companies, Lowes and Home Depot. In 2010, the health care company Kaiser Permanente installed its sustainability scorecard to evaluate a product's sustainability. This set the stage for a greener hospital movement in the health care industry, as many hospitals now require medical product suppliers to provide more sustainable options. Eco-innovation is a value driver, and instructions on how to run an eco-innovation session are presented. Product rating systems like Environmental Working Group (EWG) and YUKA for determining sustainability performance continue to emerge in the market and are discussed.

CONSUMER DEMAND FOR GREENER PRODUCTS

A 2022 joint study conducted by McKinsey and Nielsen IQ analyzed US sales data over five years and found that products with **ESG-related claims observed an average of 28% cumulative growth**, whereas products without such claims averaged 20% cumulative growth from 2017 to 2022. From their collaborative work, research showed that certain demographics, like higher-income households or households in suburban areas, are more likely to purchase products with more than one ESG-related claim. Yet, regardless of income or geographic location, Nielsen IQ's household panel exemplifies that consumers across different demographics are willing to purchase sustainable products. The findings disprove the widely held belief that greener products are demanded by a fixed target market, where in reality, consumers of all demographics want to purchase products with these attributes. The study saw that two-thirds of product categories that made ESG-related claims grew faster than those that didn't! Furthering the evidence of interest in greener products it was noted that labels like "vegan" and "carbon-zero" have higher growth rates, specifically

DOI: 10.1201/9781003441939-3

8.5% more than more typical claims and labels. Products considered in the middle tier of the claims, like "sustainable packaging" and "plant-based," had a 4.5% greater growth rate than others. Products with widespread claims still showed a 2% growth rate than those without any form of ESG labels (McKinsey & Nielsen 2023).

Topping the list, however, of sustainability purchasing drivers for a majority of consumers was a single factor: **brand trust**. In Shelton Group's 2021 survey, respondents were asked to state a "bad company" and what makes them a "bad company," and 16% of customers chose as their reason "**cannot trust/deceptive/dishonest/manipulative**." Further analysis indicates that not following ESG initiatives and having issues with "employee treatment, dishonesty/fraud/scandal, values/stands, environmental harm" will negatively impact business, getting your number one on the so-called "**Bad List**" (*Good Company* 2021). Another study supports this sentiment, indicating that **36% of Americans chose specific brands or stopped purchasing a product due to the social and environmental reputation of the company**; 26% of Americans can even provide the company's name (*Shoptivism* 2021).

> Corporate social responsibility is a hard-edged business decision. Not because it is a nice thing to do or because people are forcing us to do it … **because it is good for our business**.
>
> **Niall Fitzgerald**
> *Former CEO of Unilever*

Procter & Gamble is a prime example of consumer values translating into sales. Despite being the world's largest consumer products company, Procter & Gamble must be mindful of more sustainable brands that can erode their market share like companies with products that are seen as more natural, environmentally sensitive, or purpose-driven, such as Method, Honest Co., and Seventh Generation. Innovative sustainability minded thinking brings new competition like Kudos creating a 100% cotton disposable diaper challenging Procter & Gamble's product line with a more sustainable alternative, compared to Pampers, made of polyester and polypropylene (Segran 2022). To maintain their leadership position, Procter & Gamble will have to focus on making many of its own brands greener to appeal to sustainability-minded shoppers.

RETAILER'S DEMANDS

The greatest driver for developing greener products is when the marketplace demands it. When your customer is asking you for products that have lower environmental impacts, you pay attention. This is especially true when it's your largest customer.

In 2005, Lee Scott, at the time CEO of Walmart Stores, Inc., gave a speech—the first of its kind—to be live cast to all the company's stores. "If we were a country, we would be the 20th largest in the world. If Walmart were a city, we would be the fifth largest in America … **What if we used our size and resources to make this country and this earth an even better place for all of us**: customers, associates, our children, and generations unborn?" Thus began Walmart's sustainability journey (Makower 2015a).

When **Walmart**, the largest retailer in the world, embarked on an aggressive sustainability program and asked their suppliers to help by providing greener products, it changed the way business is conducted. Whatever the reasons are for Walmart's sustainability initiatives, it has been hugely impactful. **I cannot think of a more significant single event that propelled the development of greener products than when Walmart decided to embrace sustainability.** When the biggest retailer in the world puts out a sustainability scorecard that can help (or hurt) your sales, you stand up and pay attention.

The company set out a three-step plan: (1) develop a supplier sustainability assessment, (2) develop a life-cycle analysis database, and (3) develop a simple tool that customers can use to consume in a more sustainable way (Makower 2016a).

WALMART SUSTAINABILITY GOALS

Climate and renewable energy leadership
Zero waste in operations, products, and packaging
Regeneration of natural resources: forests, land and oceans
Sustainable product supply chains
Dignity of people in supply chain

Walmart ESG Report 2022

If you sell to Walmart, you will be asked to help provide sustainable products, reduce greenhouse gas emissions, and help them to produce zero waste. Part of their first steps in obtaining sustainability information on products sold in their stores is through their supplier sustainability assessment. This is a survey consisting of 15 questions that addressed energy and climate, material efficiency, natural resources, and people and community. The questions that were asked helped to spark change in the industry, which in no doubt had companies scrambling, forcing action in fear of competitors having an advantage. Consider the implications of the following questions; if your answer is <u>no</u> and your competitor has programs in place, this can put you at a competitive disadvantage.

- Have you measured and taken steps to reduce your corporate greenhouse gas emissions? If yes, what are those targets?
- Have you set publicly available water-use reduction targets? If yes, what are those targets?
- Have you obtained third-party certifications for any of the products that you sell to Walmart?
- Do you invest in community development activities in the markets you source from and/or operate within?

(Walmart 2016)

In 2017, Walmart introduced Project Gigaton to support the Paris Agreement 2° Warming Scenario, whose main objective is to contribute to limiting global warming to no more than 2°C. The initiative is set to encourage Walmart's suppliers to reduce one gigaton of greenhouse emissions by 2030. Suppliers are invited to gauge their own contributions to the project and participate in one or more of the pillars: Energy, Nature, Waste, Packaging, Product Use, and Transportation. Walmart welcomes goals of all sizes, but encourages their goals to be SMART: specific, measurable, achievable, relevant, and time limited. The company strategically worked with the CDP (a not-for-profit charity), the Environmental Defense Fund (EDF), and the World Wildlife Fund (WWF) in creating science-based guidelines, allowing suppliers to report their own CO_2 data. As of 2022, more than 4,500 suppliers participated in Project Gigaton, where greater than 750 million metric tons of cumulative emissions have been reduced or avoided since 2017, at 75% of their goal (*Project Gigaton* 2023)!

The company has been extremely successful by several measures establishing itself as a leading force. By the end of 2021, Walmart reduced combined scopes one and two emissions by 17.5%, diverted 78% of waste globally, and has approximately 58% of global private brand packaging to be recyclable, reusable, or industrially compostable (Walmart ESG Report 2022). Since Lee Scott set the first sustainability goals for Walmart, the company continues to up their game with a new set of goals for the future. The company plans to:

- Target zero emissions in their own operations by 2040.
- Reach 100% renewable energy by 2035.
- Work with suppliers to avoid 1 gigaton of greenhouse gases emission from the global value chain by 2030.
- Protect, manage or restore at least 50 million acres of land and 1 million square miles of ocean by 2030.
- Source at least 20 commodities more sustainably by 2025.
- Achieve 100% recyclable, reusable, or industrially compostable private-brand packaging by 2025.
- Achieve zero waste in their operations in the U.S. and Canada by 2025.

(*Walmart ESG Commitments, 2023*)

When you consider the quantity and diversity of products that are sold in their stores around the world, there has been a profound impact on their supply chain. Besides typical consumer brands, there are gardening products, pharmacy products, eye care, home furnishing, electronics, and much more. All their suppliers are being forced to consider the sustainability of their products in a way they may have never had to. I can tell you from my experience working with Johnson & Johnson, we were well aware that whenever we had a meeting with Walmart, we needed to be prepared to discuss what our brands are doing about sustainability.

All this is completely voluntary and is a business-to-business (B2B) sustainability initiative—a true market driver!

COSMETIC RETAILERS

Owned by LVMH, **Sephora** is considered one of the biggest international beauty retailers providing a wide selection of beauty brands. With over 2,700 stores in 35 countries worldwide and over 46,000 employees, this growing business continues to take charge of the beauty industry, surely having a significant influence worldwide (*Sephora About Us* 2023). As of 2022, the company's stores, distribution centers, and headquarters are powered by 100% renewable energy and have been an Environmental Protection Agency (EPA) Green Power Partner for the last seven years (*Sephora* Sustainability 2023). In 2021, Sephora diverted more than 1.6 million pounds of waste and continues looking for more ways to grow their number by partnering with Pact, a nonprofit group dedicated to recycling empty beauty packaging (Pact Collective 2023).

As part of the French multinational beauty conglomerate, Sephora will increase its efforts in sustainability to help achieve LVMH's goals. In their LIFE 360 Program, they set the following goals:

- Restore and/or regenerate 5 million hectares of flora and fauna habitat by 2030.
- 100% of new products covered by a sustainable design approach by 2030.
- 100% of strategic supply chains covered by a dedicated traceability design by 2030.
- 55% reduction and/or avoidance of Scope 3 greenhouse gas emissions by 2030.

(LVMH 2022)

In addition to their climate change efforts, Sephora is committed to providing their consumers with the best high-quality products, focusing on safety and sustainability. In 2019, Sephora launched its Public Chemicals Policy to reduce any high-priority chemicals, encourage safer replacements, be transparent with its consumers, and engage in their knowledge. In their latest report, 95.5% of Sephora's selection did not contain high-priority chemicals and they continue to focus on the remaining to achieve 100%. In addition, 100% of all intended ingredients are listed on the US Sephora website, providing customers with straightforward answers to their questions about a product's chemical formulation (*Sephora Public Chemicals Policy* 2022). Working with Estée Lauder Companies, I can say that not only Sephora but other prestige beauty retailers like ULTA Beauty, Macy's, Nordstrom and Selfridges also have sustainability targets that suppliers must be mindful of. Many of these retailers **have set clean beauty designation** for certain products; if you meet specific requirement like being free of ingredients of concern (e.g. parabens, triclosan, etc.), vegan and cruelty free. Companies are trying to meet these retailers' clean beauty requirements to give them an edge over competitors. If your company wants to sell products in a Sephora store, then it would be in your best interest to have similar sustainability objectives to them.

HOME IMPROVEMENT COMPANIES

Home improvement companies have gotten in on the greener product movement too, seeing opportunities to align their offerings with their customers' growing desire for products with environmental and social benefits. **Lowe's** has set a sustainability strategy that emphasizes bringing greener products to customers. The strategy is to provide environmentally responsible products, packaging, and services at everyday low prices. Some of the objectives include:

- Align goals and initiatives to applicable US and global standards, such as the United Nations Sustainable Development Goals.
- Contribute to environmental protection through responsible sourcing policies.
- Refine the supply chain to operate more efficiently, including the reduction of waste and carbon footprint.
- Increase transparency around the environmental impacts of operations, including measuring greenhouse gas emissions.

(Lowe's Sustainability Policy 2023)

It is obvious that suppliers offering products with improved performance will get preference in Lowe's stores. Lowe's reports progress based on the environmental benefits from products sold in the form of energy and water savings. In 2021, Lowe's sold enough Energy Star products to:

- Save consumers $23.5 billion yearly on their energy bills compared with non-Energy Star-qualified products since 2018. By 2025, the company plans to reach $40 billion in savings.

The number of WaterSense-labeled toilets and bathroom faucets sold in 2021 can save enough water in a year to:

- Save 1 trillion gallons of water annually.
- Save consumers $11.5 billion in lifetime water bills.

(Lowe's Corporate Responsibility Report 2021)

Now that's an amazing reduction in environmental impacts and dollars saved through the sale of greener products!

Their competitor, **Home Depot**, has also set goals to bring greener products to their customers. One objective is to encourage customers to become environmentally conscious shoppers. In 2007, Home Depot launched the at-a-glance **Eco Options** identification system, which helps educate consumers about products with improved environmental performance. In 2019, Home Depot offered over 25,000 Eco Options products (Home Depot 2019). A product is classified as having improved performance if it demonstrates benefits in one of five areas: **energy efficiency,**

water conservation, healthy home, clean air, and **sustainable forests.** Improvements are judged by third-party certifications that have been given to products like the USEPA's Design for Environment, USDA Organic, Forest Stewardship Council, USEPA Energy Star, and other criteria (*Eco Options* 2023). As we will see in the green marketing section of this book, third-party certifications do build trust and confidence in claims made by companies, so this is a wise approach.

The type of products that are given the Eco Options designation include low VOC paint, WaterSense®-labeled bathroom fixtures, Energy Star®-labeled electrical products, organic plant food, and environmentally preferred cleaners. Any company wishing to sell products in a Home Depot store will need to pay attention to their Eco Options program and seek this designation. In addition to the improved environmental performance, these greener products give customers cost savings, which Home Depot makes sure to highlight. The Eco Option program is a good way for manufacturers to differentiate their products and delight customers.

GREENER HOSPITALS

Companies providing products to hospitals are not exempt from the greener product revolution. Why would hospitals be demanding the development of greener products? Consider that they are operating 24/7, the lights are always on, waste is constantly generated, disinfectants and various other chemicals are being used, air pollutants are generated by boilers that supply heat and hot water, and wastewater is continually flowing. There are financial benefits for reducing a hospital's footprint—according to Kathy Gerwig, former VP and Environmental Stewardship Officer for Kaiser Permanente, their Environmentally Preferred Purchasing program resulted in \$63 million in annual savings from reducing energy, waste, and toxic chemicals in 2015 (Gerwig 2015).

Hospitals are being encouraged to become more sustainable by interest groups like Health Care without Harm. Using a phrase from the Hippocratic Oath that doctors take, their mission is to see health care "**first, do no harm.**" Their goal is to encourage health care providers to do away with practices that harm people and the environment. The link between human health and environmental pollution is a point used to enroll more hospitals. Product manufacturers are impacted by this movement since there is a focus on purchase of safer products, materials, and chemicals. Hospitals are trying to avoid products containing toxic materials such as mercury, polyvinyl chloride (PVC) plastic, and brominated flame retardants (*Health Care without Harm About Us* 2023). In 2023, the Non-Governmental Organization (NGO) Health Care without Harm became an official health care sector partner of Race to Zero, a UN supported global campaign committed to achieve net-zero greenhouse gas emissions. They have announced that over 70 health care companies, with more than 14,000 hospitals from 25 countries, have joined the campaign (*New Race to Zero* 2023).

One of the leading health care providers, **Kaiser Permanente**, raised the bar for greener health care products when it unveiled its sustainability scorecard in 2010. Each company intending to sell to Kaiser Permanente is to complete the scorecard, and the results will be used to make purchase decisions. I have heard it stated at public conferences that a product's sustainability performance can be up to 20% of the purchase decision depending on product category. So, companies need to take their scorecard seriously if they want to sell to Kaiser. The focus areas for the company are the use of toxic substances; an example would be understanding if a product contains Di(2-Ethylhexyl) Phthalate or DEHP. This chemical is undesirable because it has shown adverse effects on the development of the male reproductive system in young laboratory animals, and there is some concern this could also affect some human patients. So, one question on Kaiser Permanente's scorecard is: "Does the product contain DEHP?" If the answer is "no," the vendor enters a 0; if the answer is "yes," it scores 1. The lower-scored products are the more environmentally friendly. The implementation of this innovative scorecard can impact how medical-device manufacturers do business since Kaiser Permanente purchases more than $1 billion each year of medical products (Hicks 2010).

In June 2016, Kaiser Permanente announced their new "Green Goals" for 2025. These targets as well as their current status are as follows:

- Reduce water by 25% per square foot—A The company reduced its water use by 15% between 2013–2021.
- Recycle, reuse, and compost 100% of non-hazardous waste—In 2021, Kaiser Permanente diverted 51% of its non-hazardous waste from landfill.
- Become "carbon net positive" by buying enough clean energy and carbon offsets to remove more greenhouse gases from the atmosphere than it emits—In 2020, the company achieved carbon neutrality through the usage of renewable energy sources and energy-efficient use.
- Buy all of its food locally or from farms and producers that use sustainable practices, including the use of antibiotics responsibly—In 2019, 43% of the food purchased comes from farms and other producers using sustainable practices.
- Increase its purchase of products and materials that meet environmental standards to 50%—The company currently eliminated the following chemicals from products: polyvinyl chloride (PVC), diethylhexyl phthalate (DEHP), triclosan, triclocarban, and fluorochemical additives.

(Kaiser Permanente 2021)

To achieve these goals, Kaiser Permanente will have to rely heavily on their suppliers providing greener products. With their large purchasing power, the health care industry is an important driver of more sustainable products.

B2B PURCHASING

Greener products are not only relevant to consumers, we have also seen a strong pull from B2B customers. The phrase, "greening the supply chain," has become synonymous with one business asking another to become more sustainable. Companies are pressured to become more sustainable on many fronts. One area that was not originally foreseen was the focus on the supply chain, from procurement of services and raw materials to third-party manufacturers.

More companies are asking their suppliers to help them with their sustainability goals. Unilever has been pressured for using palm oil and other agricultural raw materials in their products that are sourced from farms that have been accused of damaging tropical rain forests. In 2020, Unilever introduced the **People and Nature Policy**, focusing on sourcing and being more selective in employing suppliers who advocate for people and the planet (Unilever 2023). Walmart set goals for its suppliers in China to reduce or avoid greenhouse gas emissions by 50 million megatons by 2030 (Walmart 2018). In 2022, Staples asked more than 300 suppliers to complete their new sustainability supplier assessment to help better evaluate the company's supply chain (Staples 2023). SC Johnson, in their efforts to minimize impact on the environment and support universal human rights, developed the **SC Johnson Supplier Code of Conduct**, which specified the minimum requirements that all SC Johnson suppliers are required to comply. The intent for this initiative is that their suppliers will do "what's right" (SC Johnson 2023).

> We are guided by our principles, and we expect SC Johnson suppliers to share this commitment to doing what's right.
>
> **Fisk Johnson**
> *Chairman and CEO of SC Johnson (SC Johnson 2023)*

In 2010, **Procter & Gamble** (P&G) developed a **supplier Environmental Sustainability Scorecard**. The scorecard was implemented to determine which suppliers P&G would conduct business with based on the overall ratings of suppliers. One of the goals of the scorecard was to encourage suppliers to implement more sustainability initiatives. As a requirement, suppliers needed to provide data on their sustainability efforts like electrical and fuel use, water input and output, Scope 1 and Scope 2 greenhouse gas emissions, waste sent to landfill or incinerated, and hazardous waste disposal in 2016 (Procter & Gamble 2016). P&G believed the rating system would encourage environmental improvement across its entire supply chain at the time of 2016, since they had approximately 75,000 suppliers resulting in about $65 billion in annual spending (Procter & Gamble 2016). The scorecard helped to propel sustainability in P&G, remaining competitive with other firms. These are just a few examples of how businesses are looking to their suppliers to help with their sustainability initiatives and to help them green up their products. Companies are no longer only being asked to be responsible for their own footprint but are being held accountable for addressing their entire supply chain. **Suppliers with a good sustainability story can gain an edge with their key customers.**

ECO-INNOVATION AS A VALUE DRIVER

More companies are seeing sustainability as a way to drive innovation to generate new products. There are several ways sustainability can add value to your brand.

Communicating product attributes that emphasize a brand's sustainability is an easy way to get some quick wins.

Addressing a growing apprehension to use plastic bottled water, **Brita water filters** (a Clorox product) initiated a repositioning of its product. Brita took advantage of consumers' growing desire to become greener and, what I call the **war on plastic**, by highlighting the use of their product as more sustainable than bottled water. They calculated the millions of empty bottles that would be taken out of the waste stream to make the case that their product is the more sustainable option for getting clean drinking water. Every Brita filter used saves 900 standard 16.9 oz. water bottles from being used and discarded (Brita 2023). In 2022, Brita customers replaced over 10.5 billion plastic water bottles when using the filters (*Brita Drink Like You Care* 2023). The overall product positioning is all about how a customer can help the environment by using their product.

If you go to the Brita website, you will see this very clearly, under "Why Brita": "**Better Water, Better Value, Better World, and Better Health**." In other words, you protect the environment, save money, and live a healthier lifestyle at the same time! A single Brita pitcher with filters can save consumers $325 per year with a Brita Elite filter than buying bottled water (Brita 2023). The Brita water filter repositioning is an excellent example of using innovative thinking to uncover existing sustainability attributes of your product without having to do any physical changes to the product itself.

GE's Ecomagination provided another example of repositioning products by emphasizing their existing greener attributes. As an early adopter of environmental and greener products, GE positioned and marketed the attributes under the Ecomagination banner. GE's CEO at the time, Jeff Immelt, realized that they were selling windmills, had more efficient locomotive and jet engines, and were recognized as a US EPA Energy Star partner, which sparked his epiphany: "Maybe there is something there if we put all of those together" (Makower 2009). Consider washing machines that use less water than competitors (or previous versions), hot water heaters and microwave ovens that use less energy—all of these can be positioned under the Ecomagination banner. The reason why the equipment has better environmental performance doesn't matter; the key is *if it does*, **communicate it to your customers!**

Use of sustainability as the driver for developing new product concepts is another way to meet market needs.

A good example of this is Samsung's greener product development process, used to generate innovative new products that have substantial sustainability benefits. In 2022, Samsung designed the new Galaxy Z series, Galaxy Buds2 Pro, and other products with repurposed fishnets, eliminating possible waste from the ocean, as

640,000 tons of fishing nets are abandoned in Earth's waters every year. Repurposed fishnets, post-consumer materials, or bio-based resin composed 90% of Samsung products launched in 2021. The Galaxy 22 and the Galaxy Z series utilized 100% of recycled paper for packaging, saving approximately 51,000 trees (Samsung 2022).

Another example of sustainability as an innovation driver is The Clorox Company's line of natural cleaners, EcoClean™. Clorox's cleaning and disinfecting products are not particularly known for being environmentally friendly. However, the company utilized sustainability as the basis and developed the brand of cleaning products. The Clorox EcoClean follows the EPA Safer Choice and Better for the Environment program standards. The product is made from a plant-based active ingredient, and it is available in 25% post-consumer recycled plastic (The Clorox Company 2023).

The electric vehicle (EV) manufacturer, Tesla sparked eco-innovation in the automotive industry in a major way with its commitment to a sustainable future by "building a world powered by solar energy, running on batteries and transported by electric vehicles" (*Tesla About Us* 2023). Since the launch of the Model S in 2012, Tesla's ongoing commitment to innovation led the company to create an unmatched electric vehicle (Barry 2022). Tesla delivered 1.31 million cars in 2022, a record high for the company, with a 40% increase from 2021's sales (*Tesla Vehicle Production* 2024). With the company taking hold of the market, competitors are joining the EV race looking to erode Tesla's market share. General Motors and Toyota have launched affordable electric vehicles providing greener products for a larger population. The US government struck a deal with Tesla, where the company will create 7,500 open-access chargers by 2024. As part of the Biden Administration $7.5 billion campaign to fund 500,000 chargers across America (Mitchell 2023). The movement is not only pushing for the US government to make efforts, but Japan has also stated that all new cars sold must be environmentally friendly by 2035 (International Trade Administration 2021).

In 2022, Beiersdorf introduced the expansion of its "Climate Care" strategy at the product level, hoping to achieve climate neutrality for its *NIVEA Core Assortment*. The company's bestseller *NIVEA SOFT* will undergo a greener upgrade, with the new formulation being 98% biodegradable, reducing approximately 40% of carbon emissions in the product's footprint. A further commitment to eco-innovation, Beiersdorf announced that they will be the first firm to use Carbon Capture Utilization (CCU) Technology, a process in which carbon dioxide emissions are captured and chemically altered into products like the company's new *NIVEA MEN Climate Care Moisturizer*. In addition, the formulation is free of undesirable ingredients like microplastics, silicones, mineral oils, and polyethylene glycol (PEG) and its derivatives. Beiersdorf is taking bold steps in the beauty sector to create greener products through its commitment to greener formulas and reducing carbon dioxide emissions (Beiersdorf 2022).

It is common for companies to run innovation sessions to develop new product concepts. As we have seen, many companies have been successful in using sustainability as the innovation driving force. I have had the opportunity to run some **eco-innovation sessions** and have been surprised to see that top executives are unfamiliar with the sustainability achievements that their business unit has already accomplished. If the executives are unaware, our customers are as well. With the growing demands for more sustainable products in all types of businesses, it is a wasted opportunity to not communicate the improvements made already and to generate new product concepts.

ECO-INNOVATION

Eco-innovation is a value driver, and companies are continuously searching for ways to make improvements of their products sustainability.

ECO-INNOVATION SESSION AGENDA

Eco-innovation doesn't have to be a complicated thing. Simple tools can be used to spur innovative thinking. Some of the key groups that you would want to include in an innovation session would be marketing, sales, R&D, operations, procurement, environment, health & safety, and communications. A simple agenda for one of these events would look like this.

- *Landscape*: What's going on in the marketplace? Are customers looking for greener brands?
- *Competitor Analysis*: Is anyone leading in sustainability? Are we competing against any green brands? Do any competitors have weaknesses compared to our brand?
- *Company Accomplishments*: What have we accomplished as an enterprise and are there any specific greener attributes to our brand (e.g., use of recycled content, better performance, end-of-life solutions)?
- *Risk Analysis*: What happens if we do nothing? Do we have any sustainability problems (perceived or actual)?
- *Opportunity Analysis:* (1) What can we communicate to customers that we have accomplished already? and (2) what can we do to green up our brand or develop a new product?
- *Prioritization*: Develop a prioritized list of projects and assignments (Unruh & Ettenson 2010).

If developing a new product concept or greening up an existing brand is a desired outcome of the innovation meeting, then the Eco-design elements in Figure 2.1 are a good way to initiate innovative thinking.

Just taking a quick look at Figure 2.1 can spark many sustainable product innovations. If customers have been complaining about excessive packaging or that it can't be recycled, then we should initiate a project to reduce the packaging size and use commonly recycled materials. **Once the improvement is accomplished, make sure you tell them about it.** The engineering department mentioned that we reduced the energy use of the product by 10% in its latest design and we think we are best in class for energy use—we should do a competitor analysis and let our customer know that our product is saving them money and is better for the environment. We are losing some share points to a competitor with a new natural brand; let's develop our own

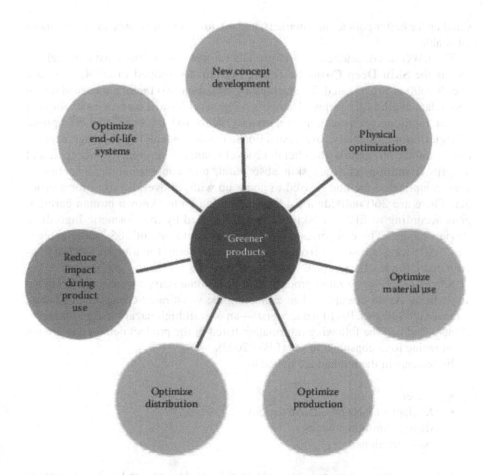

FIGURE 2.1 Eco-design elements. (From H. Brezet and C. Hemel, PROMISE Manual of Ecodesign, adapted by Natural Resource Council Canada, 2007, Design for Environment Guide. Five Winds International. Ottawa, Canada, p. 19.)

natural product line or try to "green up" by removing ingredients that the marketplace is concerned with.

PRODUCT RATING SYSTEMS

Another reason to be mindful of a product's sustainability is that there are many independent groups that are rating your product and making it very easy to find out the score. "There's no room for green-washing anymore," states Chris Sayner from specialty chemicals company Croda. "**People just won't accept your word for it: independent verification is now necessary to give true transparency**" (Whitehouse 2016). EWG is an NGO with a product-rating system, with six major program areas: toxics, food, agriculture, children's health, energy, and water. EWG's mission is threefold: **advocate** for a healthier, greener world, **activate** consumers to stand up for better policies and markets, and **advance** sound science in environmental health.

The EWG has developed several product ratings. One of the most comprehensive is the **Skin Deep Cosmetic rating system**. Developed in 2004, there are over 90,000 products rated. The ratings are based on two factors: a hazard rating and a data availability rating. There are 17 general hazard categories evaluated: cancer, reproductive/developmental toxicity, neurotoxicity, endocrine disruption, allergies/immunotoxicity, use restrictions, organ system toxicity, persistence/ bioaccumulation, cellular/biochemical level changes, ecotoxicity, occupational hazards, irritation, enhance skin absorption, and contamination concerns. A very complicated method is used to come up with an overall rating for a product. There are 260 individual categories ranging from "known human carcinogens according to EPA" to "skin irritants identified by the Cosmetic Ingredient Review panel." The categories are then mapped into one of 205 "score categories." All scores are scaled from 0 to 10, with 10 being the highest level of concern and 0 the least.

Some of their worst-rated products can have some scary sounding information about them. As an example, when evaluating the worst-rated cosmetic foundation, it received a score of 10 (10 being worst)—an overall high hazard listing for ingredients—and had the following information listed in the product details that are not encouraging for a consumer to see (EWG 2023).

Ingredients in the product are linked to:

• Cancer
• Developmental/reproductive toxicity
• Allergies/immunotoxicity
• Use restrictions

Other concerns for ingredients used in this product are endocrine disruption, persistence and bioaccumulation, non-reproduction organ system toxicity, multiple additive exposure sources, and contamination sources (aniline and cadmium). Other than that, it is a pretty good product! Any customer or potential customer that happens to review this database would, at the very least, question their purchase of this product.

In addition to the ratings they provide, EWG has launched their **EWG VERIFIED®** program, which highlights products that go above and beyond its green rating in EWG's Skin Deep because the company has disclosed more about its formulations and manufacturing processes. To become EWG VERIFIED®, a product must meet all of EWG's strictest criteria, including:

• Products must fall into one of the EWG personal care product categories approved for licensing.
• Products must score a "green" in EWG's Skin Deep database.
• Products cannot contain any ingredients on EWG's "unacceptable" list, meaning ingredients with health, eco-toxicity, and/or contamination concerns.

- Products cannot contain any ingredients on EWG's "restricted" list, which do not meet the restrictions set by authoritative bodies and industry institutions.
- Products must follow standard ingredient naming guidelines.
- Products must fully disclose all ingredients on the label, including ingredients used in fragrance.
- Product manufacturers must develop and follow current good manufacturing practices.
- Products must follow the European Union requirements for labeling fragrance allergens.
- Products must follow the European Union labeling guidelines for nanomaterials used in cosmetics.
- Product labels must indicate an expiration date or a "period of time after opening."
- Products must pass basic microbial challenge tests and repeat these tests as appropriate.

The EWG VERIFIED® program makes it easy for consumers to identify the highest-rated products. Walmart has teamed up with EWG VERIFIED® to launch its "Build for Better" program helping consumers shop for healthier and sustainable products (*Walmart's Built for Better* 2021). "Build for Better" includes products that have received EWG verification, allowing customers direct access to such initiatives.

Amazon collaborated with outside certification groups to create the Climate Pledge Friendly Program in helping customers discover and shop for approved sustainable products. Another product rating program is **Amazons Climate Pledge Friendly Program**. Since 2020, Amazon has over 250,000 products included in this program, and in 2021, more than 370 million units of products with this designation were shipped to Amazon customers (Amazon 2023).

In Europe the Yuka rating was founded in France in 2017. Yuka helps consumers choose the right food and personal care products for them (Southey 2019). The Yuka app is taking hold of Europe with more than 30 million downloads and 6 million active users.

Regardless of how valid a company may feel these ratings or labeling systems are, it merits to pay attention to them and determine whether there is something to be learned that could influence product improvements. I can tell you that at the Estée Lauder Companies Inc. we look at how our products are rated and consider the ingredients that are rated low to evaluate what the rating organizations are stating is of concern; the key here is to consider criticism of your products and evaluate the concern to position your products in the best possible way.

ESG FUNDS

Another trend that is encouraging the development of more sustainable products is socially responsible investing. More investors are considering issues beyond corporate profit and are interested in purchasing stock only from companies that align with their personal values. ESG funds help to invest in companies taking the initiative in having a sustainable and positive societal impact in the world (Benson 2023).

In the corporate world, companies tend to view ESG-related spending as a cost rather than an investment (Poleman & Winston 2022). However, there is an overwhelming amount of evidence that debunks such theories. JUST Capital is a nonprofit research organization providing society's stakeholders with information on their assessment of JUST companies. In 2021, the firm released the "2021 Rankings of America's Most JUST Companies" and found that the highest rated businesses had the best returns on equity (Mahoney 2020)! The financial services company, Refnitiv Lipper, states in 2021 "a record $649 billion poured into ESG-focused funds worldwide...up from the $542 billion and $285 billion that flowed into these funds in 2020 and 2019" (Jessop & Kerber 2021). Additionally, Bloomberg Intelligence suggests global ESG assets are on track to surpass $53 trillion by 2025 (Bloomberg Intelligence 2021). It is a pretty good bet that these funds will continue to expand, and it is a wise thing for publicly traded companies to pay attention to how their company is being perceived by these funds.

Listed below are the Best-Performing ESG funds as of 2023:

- Amana Growth Investor
- Empower Large Cap Growth Inv.
- Brown Advisory Sustainable Growth Inv.
- Neuberger Berman Large Cap Growth Inv.
- VALIC Company I Large Capital Growth.
- American Century Select Inv.
- Commerce Growth.

(Benson 2023)

ESG RATERS/RANKERS

Rating and ranking systems are prominent metrics that encourage socially responsible investing. These are things that company top management pay attention to; a good indicator of if your firm is in a leadership position for sustainability. The consulting group, Environmental Resource Management (ERM) Group, conducted research and discovered **"investors often rank ESG ratings above a company's own disclosures as a useful source of information on corporate ESG performance"** (Framework ESG 2021). Some key ESG rating agencies include:

- Morgan Stanley Capital International (MSCI)
- Institutional Shareholder Service Inc., ISS ESG
- Bloomberg
- Sustainalytics
- Vigeo Eiris, Moody's ESG Solutions
- Dow Jones Sustainability Indexes (DJSI)
- CDP Disclosure Insight Action
- FTSE4Good

(Framework ESG 2021)

There are many rating and ranking agencies, and each have their own "methodology, scoring and target audience." For example, the CDP rates companies based on the information voluntarily disclosed and reported in the questionnaire pertaining to the following three programs: climate change, water, and forests. The DJSI has a detailed questionnaire that evaluates the sustainability performance of companies, considering "corporate governance, risk management, branding, climate change mitigation, supply chain standards, and labor practices" (Framework ESG 2021).

In order to do well in these indexes, a company must be high performing in all aspects of sustainability, including bringing greener products to market. More focus is being placed on product stewardship, eco-efficiency, and supply-chain issues associated with product manufacturing. Therefore, failure to adequately address product stewardship issues like potentially toxic materials in your product, not adequately addressing product end-of-life issues, or failing to initiate sustainability programs at your suppliers, can hurt your rating in these indexes.

SHAREHOLDER RESOLUTIONS

In addition to social responsibility funds and investing pressures, we have seen an increasing amount of **shareholder resolutions** focused on product responsibility. According to the Proxy Prereview Project, "the 2022 U.S proxy season has yielded a record 607 shareholder resolutions on environmental, social and governance issues, a 22% increase from 2021" (Rives 2022). Using shareholder resolutions is a smart strategy which environmental groups have initiated. In the United States, a shareholder of a publicly traded company can propose a resolution which is to be voted on by all shareholders. Therefore, environmental oriented NGOs buy stock in companies they want to influence. Based on my experience I can tell you that companies go through great lengths to try and address the concerns of environmental groups to prevent these resolutions from going to a shareholder vote fearing adverse public perception of their company.

An example of how the resolution process is used would be activist investor company Engine No.1 set a bold campaign against Exxon Mobil Corporation demanding the company take initiative on carbon neutrality and the replacement of four directors from the board (Phillips 2021). Engine No.1 backed by other shareholders of the company successfully overthrew at least two directors marking an unprecedented win against the energy company. Similarly, the shareholders of Chevron Corp. voted in favor of a proposal to help cut carbon emissions and reduce the company's carbon footprint in 2021 (Reuters 2021). In the cases of Exxon and Chevron Corp., shareholders believed that the companies were not fulfilling their duties when it came to protecting the environment and forced the company to take different actions.

In the fast-food industry, activist investor, Carl Icahn, called out McDonald's lack of animal welfare and pushed for the company to achieve its mission of being crate free. Mr. Icahn advocated for two ESG experts as board members to help move the company away from suppliers using gestation crates which are considered inhumane by some, by the end of 2023 (Torrella 2022). The Humane Society, an animal welfare organization, also added to Icahn's statements and called out McDonald's on its continued usage of gestation crates. The institution filed a complaint to the US Securities and Exchange Commission in attempts to hold McDonald's accountable in their treatment of livestock (Block 2022). Although Carl Icahn lost the proxy battle with McDonald's, it is important to note that companies will continue to have such conflicts on ESG initiatives via shareholder resolutions and these initiatives make companies take notice and change direction.

COMMON SHAREHOLDER PRODUCT STEWARDSHIP RESOLUTIONS

- Sustainable sourcing of raw materials.
- Addressing climate change.
- Use of GMO materials.
- Reducing toxic chemical usage.
- Addressing plastic packaging.

GREEN PUBLIC PROCUREMENT

A significant new approach to procurement is the purchase of goods and services that foster lower environmental impact. This Green Public Procurement (GPP) is a process whereby public authorities seek to procure goods, services, and works with a reduced environmental impact throughout their life cycle.

In Europe they have realized that green purchasing can have a big impact. Every year, the European Union public authorities (central, regional, and local levels) spent approximately 14% of EU GDP—or €1.8 trillion—on goods, services, and works. A good deal of this spending is on services that have significant environmental impact, such as transportation, buildings, and food. By using GPP criteria, public officials can sway the purchase of items to reduce their impact (European Commission 2023). European Union member states have developed National Action Plans (NAPs) for greening their public procurement. The NAPs contain targets that are reported on publicly. The type of goods and services that are of primary focus includes paper, cleaning chemicals, office IT equipment, construction, transport, furniture, electricity, windows, doors, thermal insulation, road construction, and mobile phones (Otero Matias 2023). Any company that manufactures products in these categories should be interested in evaluating the criteria that the EU has set for GPP.

In the United States, the EPA has been tasked with the development of Environmentally Preferable Purchasing (EPP) guidance for federal agencies to implement. Similar to the initiative in Europe, the impact can be significant. "The Federal government is the single largest purchaser in the United States, spending over $630 billion each year on a wide variety of products and services" (USEPA 2023c). This purchase power carries with it a significant environmental footprint.

The federal government can minimize environmental impact and spending cost through the purchase of goods and services with lowest impacts. The EPA stated, "In 2020 alone, the federal government purchased more than 27.4 million Electronic Product Environmental Assessment Tool (EPEAT)-registered products, resulting in a cost savings to the federal government of around $1 billion" (USEPA 2023). The EPA defined EPP as products or services that "have a lesser or reduced effect on human health and the environment when compared with competing products or services that serve the same purpose. This comparison may consider raw materials acquisition, production, manufacturing, packaging, distribution, reuse, operation, maintenance or disposal of the product or service." Guiding principles were developed to help federal agencies follow through with these commitments.

US EPA GUIDING PRINCIPLES FOR ENVIRONMENTALLY PREFERRED PURCHASING

- *Environment + Price + Performance = EPP*: Include environmental considerations as part of the normal purchasing process.
- *Pollution Prevention Emphasize*: Pollution prevention as part of the purchasing process.
- *Life-Cycle Perspective/Multiple Attributes*: Examine multiple environmental attributes throughout the product and service's life cycle.
- *Comparison of Environmental*: Impacts Compare environmental impacts when selecting products and services.
- *Environmental Performance Information*: Collect accurate and meaningful environmental information about environmental performance of products and services.

USEPA (2017)

Scorecards were developed to help federal agencies to track the progress they were making, particularly the President's Office of Management and Budget. Here again, companies that sell to the federal government need to be mindful of the potential impacts of this initiative. **Consider losing a sale to a large customer like the US government because your product was not on par with your competitors' product.**

CONCLUSION

The market is demanding greener products and this demand is growing. Manufacturers need to fully realize the growing pull for products with enhanced environmental and social benefits.

MARKET DEMANDS FOR GREENER PRODUCTS

- Consumers desire greener products
- Institutional customers are requesting them
- Hospital green revolution
- B2B green purchasing
- Growth of ESG Stock Funds & Sustaiability Ratings
- Green Public Procurement
- Eco-innovation as a value driver

No matter what aspect of business you are in, there is a shift occurring that makes it an imperative to offer greener products to your customers. We are seeing market demands in all major sectors: apparel, consumer goods, to chemicals, transportation, medical products, pharmaceuticals, energy, petroleum, and others. Aside from customer demand, there is pressure from other areas such as environmental groups, sustainability rating organizations, and competitors. Sustainability is being used as an innovation driver for new product development. Having greener product offerings is no longer "a nice to have" prospect, but a necessity to be competitive in the marketplace.

REFERENCES

About the Environmentally Preferable Purchasing Program. 2017. The United States Environmental Protection Agency. https://19january2017snapshot.epa.gov/greenerproducts/about-environmentally-preferable-purchasing-program_.html (Accessed November 20, 2016).

About the Environmentally Preferable Purchasing Program. 2023. The United States Environmental Protection Agency. https://www.epa.gov/greenerproducts/about-environmentally-preferable-purchasing-program#:~:text=In%202020%20alone%2C%20the%20federal,government%20of%20around%20%241%20billion (Accessed July 10, 2023).

About Us. 2023. Health Care without Harm. https://noharm-uscanada.org/content/us-canada/about-us#Mission (Accessed July 10, 2023)

About Us-Accelerating the World's Transition to Sustainable Energy. 2023. Tesla. https://www.tesla.com/about (Accessed June 9, 2023).

About Us-Sephora. 2023. Sephora. https://www.sephora.com/beauty/about-us (Accessed May 30, 2023).

Annual Report. 2016. Proctor & Gamble. https://www.annualreports.com/HostedData/AnnualReportArchive/p/NYSE_PG_2016.pdf (Accessed November 2, 2016).

Barry, K. 2022. How the Tesla Model S Changed the World. *Consumer Reports.* https://www.consumerreports.org/cars/hybrids-evs/how-the-tesla-model-s-changed-the-world-a7291465820/ (Accessed June 9, 2023).

Beauty Repurposed. 2023. Pact Collective. https://www.pactcollective.org/pact-x-sephora (Accessed May 30, 2023).

Beiersdorf expands its climate-neutralized approach to NIVEA's Core Assortment and enhances holistic "Climate Care" strategy. 2022. Beiersdorf. https://www.beiersdorf.com/newsroom/press-releases/all-press-releases/2022/04/21-beiersdorf-expands-its-climate-neutralized-approach (Accessed June 14, 2023).

Benson, A. 2023. 7 Best-Performing ESG Funds and 6 Cheapest ESG ETFs for July 2023. *Nerdwallet.* https://www.nerdwallet.com/article/investing/best-esg-funds (Accessed June 1, 2023).

Block, K. 2022. We're calling for SEC action over McDonald's deception as the company confirms cruel crate confinement. *The Humane Society.* https://blog.humanesociety.org/2022/04/were-calling-for-sec-action-over-mcdonalds-deception-as-the-company-confirms-cruel-crate-confinement.html (Accessed July 11, 2023).

Brita. 2023. Brita. https://www.brita.com/ (Accessed May 24, 2023).

Chevron investors back proposal for more emissions cuts. 2021. Reuters. https://www.reuters.com/business/energy/chevron-shareholders-approve-proposal-cut-customer-emissions-2021-05-26/ (Accessed May 28, 2023).

Clorox EcoClean. 2023. The Clorox Company. https://www.cloroxpro.com/products/clorox/clorox-ecoclean/?upc=044600602134 (Accessed May 2024).

Consumers Care about Sustainability – and Back It Up with Their Wallets. 2023. McKinsey & Nielsen. https://nielseniq.com/wp-content/uploads/sites/4/2023/02/Consumers-care-about-sustainability%E2%80%94and-back-it-up-with-their-wallets-FINAL.pdf (Accessed May 17, 2023).

Corporate Responsibility Report. 2021. Lowe's. https://corporate.lowes.com/sites/lowes-corp/files/CSR-reports/Lowes_2021_CSR_06.03.22.pdf (Accessed May 22, 2023).

Drink Like You Care. 2023. Brita. https://www.brita.com/better-world/drink-like-you-care/ (Accessed May 24, 2023).

Eco Options. 2023. The Home Depot. https://ecoactions.homedepot.com/green-products/ (Accessed May 22, 2023).

Environment FAQs. 2023. European Commission. http://ec.europa.eu/environment/gpp/faq_en.htm (Accessed May 28, 2023).

Environmental, Social, and Governance Report. 2021. Kaiser Permanente. https://about.kaiserpermanente.org/content/dam/kp/mykp/documents/reports/Kaiser-Permanente-ESG-Report_2021_ADA.pdf (Accessed May 22, 2023).

Environmental, Social and Governance Summary Report. 2022. Walmart. https://corporate.walmart.com/content/dam/corporate/documents/purpose/environmental-social-and-governance-report-archive/walmart-fy2022-esg-summary.pdf (Accessed May 20, 2023).

ESG assets may hit $53 trillion by 2025, a third of global AUM. 2021. Bloomberg Intelligence. https://www.bloomberg.com/professional/blog/esg-assets-may-hit-53-trillion-by-2025-a-third-of-global-aum/ (Accessed June 12, 2023).

ESG Commitments & Progress. 2023. Walmart. https://corporate.walmart.com/content/dam/corporate/documents/esgreport/fy2023-walmart-esg-highlights.pdf (Accessed May 17, 2023).

Gerwig, Kathy. 2015. *Greening Health Care*. Oxford University Press, New York, p. 178.

Good Company. April 2021. Shelton Group. https://storage.googleapis.com/shelton-group/2021%20Reports/Good%20CompanyReport_web.pdf (Accessed May 17, 2023).

Hicks, Jennifer. 2010. *Sustainable Scorecard to Be Used to Help Kaiser Permanente Evaluate Medical Products*. https://www.triplepundit.com/story/2012/kaiser-permanente-greens-its-supply-chain-switching-safer-iv-equipment/69336 (Accessed November 2, 2016).

Home Depot Responsibility Report. 2019. The Home Depot. https://corporate.homedepot.com/sites/default/files/2019_Responsibility%20Report_FINAL_Master_Update.png_.pdf.pdf (Accessed May 30, 2023).

Japan Transition to Electric Vehicles. 2021. International Trade Administration. https://www.trade.gov/market-intelligence/japan-transition-electric-vehicles (Accessed May 22, 2023)

Jessop, S. & Kerber, R. 2021. Analysis: How 2021 became the year of ESG investing. *Reuters*. https://www.reuters.com/markets/us/how-2021-became-year-esg-investing-2021-12-23/#:~:text=A%20record%20%24649%20billion%20poured,10%25%20of%20worldwide%20fund%20assets (Accessed July 11, 2023).

Lowe's Policy on Sustainability. 2023. Lowe's. https://corporate.lowes.com/our-responsibilities/corporate-responsibility-reports-policies/lowes-sustainability-policy (Accessed May 20, 2023).

LVMH Report. 2022. *LVMH*. https://r.lvmh-static.com/uploads/2023/03/lvmh_2022_annual-report.pdf (Accessed May 30, 2023).

Mahoney, C. 2020. Chart of the Week: JUST Businesses Have a Higher Return on Equity. *JUSTCapital*. https://justcapital.com/news/chart-of-the-week-just-businesses-have-a-higher-return-on-equity/ (Accessed June 12, 2023).

Making sense of ESG ratings and rankings. 2021. Framework ESG. https://frameworkesg.com/wp-content/uploads/2021/10/FWESG_RatingsRankings2021.pdf (Accssed June 9, 2023).

Makower, Joel. 2009. The Rise of Ratings. In J. Makower, *Strategies for the Green Economy*. McGraw Hill, New York, p. 84.

Makower, Joel. 2015a. Walmart Sustainability at 10: The Birth of a Notion. *GreenBiz*. November 16. https://www.greenbiz.com/article/walmart-sustainability-10-birth-notion (Accessed November 2, 2016).

Makower, Joel. 2015b. *Walmart Sustainability at 10: An Assessment*. November 17. https://www.greenbiz.com/article/walmart-sustainability-10-assessment (Accessed November 4, 2016).

Mitchell, R. 2023. U.S. government will pay Tesla to open its charger network to non-Tesla EVs. *Los Angeles Times*. https://www.latimes.com/business/story/2023-02-15/u-s-government-will-pay-tesla-to-open-its-charger-network-to-non-tesla-evs (Accessed June 9, 2023).

New Race to Zero Milestone: 70 Health Care Institutions from 25 countries commit to net-zero. 2023. Health Care without Harm. https://healthcareclimateaction.org/new-race-zero-milestone-70-health-care-institutions-25-countries-commit-net-zero (Accessed June 1, 2023).

New Samsung Galaxy Foldables Drive More Sustainable Future While Providing the Most Versatile Mobile Experience. 2022. Samsung. https://news.samsung.com/global/new-samsung-galaxy-foldables-drive-more-sustainable-future-while-providing-the-most-versatile-mobile-experience (Accessed May 23, 2023).

Otero Matias, J. 2023. GPP NAPS May 2023. *European Union Green Public Procurement*. https://circabc.europa.eu/ui/group/44278090-3fae-4515-bcc2-44fd57c1d0d1/library/8c778517-c451-4239-8e0c-949d64914c5b/details (Accessed July 16, 2023).

P&G, Global Partners Re-Inventing Business Partnerships. 2016. Procter & Gamble. https://news.pg.com/news-releases/news-details/2016/PG-Global-Partners-Re-Inventing-Business-Partnerships/default.aspx (Accessed November 2, 2016).

Poleman P. & Winston A. 2022. Yes, Investing in ESG Pays Off. *Harvard Business Review*. https://hbr.org/2022/04/yes-investing-in-esg-pays-off (Accessed June 12, 2023).

Phillips, M. 2021. Exxon's Board Defeat Signals the Rise of Social-Good Activists. *The New York Times*. *https://www.nytimes.com/2021/06/09/business/exxon-mobil-engine-no1-activist.html* (Accessed May 28, 2023).

Rives, K. 2022. Record number of shareholder ESG proposals in 2022 defies GOP political backlash. *S&P Global Market Intelligence*. https://www.spglobal.com/marketintelligence/en/news-insights/latest-news-headlines/record-number-of-shareholder-esg-proposals-in-2022-defies-gop-political-backlash-71181308 (Accessed May 28, 2023).

SC Johnson. Supplier Code of Conduct. https://www.scjohnson.com/en/stories/a-world-with-more-opportunity/fair-treatment-and-opportunity/sc-johnson-supplier-code-of-conduct (Accessed November 12, 2023).

Segran, E. 2022. Inside the Quest to Redesign the Diaper. *Fast Company*. https://www.fastcompany.com/90757160/inside-the-quest-to-redesign-the-diaper (Accessed May 17, 2023).

Sephora Public Chemicals Policy. 2022. Sephora. file:///Users/orakalaj/Downloads/Sephora-Public-Chemicals-Policy-3rd-Progress-Update-2022_0%20(3).pdf (Accessed May 20, 2023).

Selling Greener Products and Services to the Federal Government. 2023. United States Environmental Protection Agency. https://www.epa.gov/greenerproducts/selling-greener-products-and-services-federal-government#:~:text=Additional%20market%20research-,Overview,%E2%80%9Cgreener%E2%80%9D%20products%20and%20services (Accessed July 10, 2023).

Southey, F. 2019. Evaluating the Yuka "phenomena": How effective is the scanning app in practice? *Food Navigator Europe.* https://www.foodnavigator.com/Article/2019/08/20/Evaluating-the-Yuka-phenomenon-How-effective-is-the-scanning-app-in-practice?utm_source=copyright&utm_medium=OnSite&utm_campaign=copyright (Accessed July 11, 2023).

Shoptivism, October 2021. Shelton Group. https://f.hubspotusercontent20.net/hubfs/6711429/Reports%202021/Greenbiz_2021_Shoptivism_Report.pdf (Accessed May 17, 2023).

Skin Deep Guide to Cosmetics. 2023. Environmental Working Group (EWG) http://www.ewg.org/ (Accessed May 24, 2023).

Sustainability. 2023. Staples. https://www.staples.com/sbd/cre/noheader/about_us/corporate-responsibility/environment/#z_greener_paper%20(Accessed%20November%202,%20 2016) (Accessed July 14, 2023).

Sustainability – Sephora USA. 2023. Sephora. https://www.inside-sephora.com/en/usa/sustainability (Accessed May 22, 2023).

Sustainability: The New Consumer Spending Outlook. 2022. *NielsenIQ.* https://nielseniq.com/wp-content/uploads/sites/4/2022/10/2022-10_ESG_eBook_NIQ_FNL.pdf (Accessed May 28, 2023).

Sustainable and Deforestation-Free Palm Oil. 2023. Unilever. https://www.unilever.com/planet-and-society/protect-and-regenerate-nature/sustainable-palm-oil/ (Accessed May 22, 2023).

Sustainable Products. 2023. Amazon. https://sustainability.aboutamazon.com/environment/sustainable-products (Accessed June 5, 2023).

Tesla Vehicle Production. 2024. https://www.cnet.com/roadshow/news/tesla-reports-record-revenue-for-2022-with-1-31-million-evs-sold/ (Accessed May 12, 2024).

Torrella, K. 2022. The corporate raider taking aim at McDonald's over the treatment of pigs. *Vox.* https://www.vox.com/future-perfect/22958698/mcdonalds-icahn-pork-pigs-gestation-crates-animal-welfare (Accessed July 10, 2023).

Unruh, Gregory and Ettenson, Richard. 2010. Growing Green. *Harvard Business Review. 6.*

Walmart's Built for Better program highlights EWG VERIFIED® products. 2021. EWG. https://www.ewg.org/news-insights/news-release/2021/09/walmarts-built-better-program-highlights-ewg-verifiedtm-products (Accessed May 28, 2023).

Walmart Commits to Reduce Emissions by 50 Million Metric Tons in China. 2018. Walmart. https://corporate.walmart.com/newsroom/2018/03/29/walmart-commits-to-reduce-emissions-by-50-million-metric-tons-in-china (Accessed May 22, 2023).

Walmart Offers New Vision for the Company's Role in Society. 2016b. Walmart. November 4 http://news.walmart.com/2016/11/04/walmart-offers-new-vision-for-the-companys-role-in-society (Accessed November 26, 2016).

Walmart Supplier Sustainability Assessment: 15 Questions for Suppliers. www.walmartstores.com (Accessed November 2, 2016)

Walmart Sustainability Hub: Project Gigaton. 2023. Walmart. https://www.walmartsustainabilityhub.com/climate/project-gigaton (Accessed May 22, 2023).

Whitehouse, Lucy. 2016. "No Room for Greenwashing": The Reality of Sustainability for the Personal Care Industry. *Cosmetics.* December 15.

3 Regulatory Drivers for Greener Products

Al Iannuzzi

This chapter covers the myriad of product-based regulations as a driver for new product design, causing product development teams to anticipate and monitor these requirements. I do not present an exhaustive list of environmental product regulations, but rather a detailed review into some of the most significant ones that multinational companies have to comply with. Regulations from the European Union like the Restriction of the Use of Certain Hazardous Substances in Electrical and Electronic Equipment Regulations (the "RoHS Regulations") and global packaging restriction regulations are reviewed. Addressing these requirements affects product design criteria. To prevent barriers to the sale, product designs have to be changed years prior to the compliance date. Companies that can anticipate these regulations and make changes quicker than the competition can gain in the marketplace by appealing to purchasers that desire greener products, if they can certify compliance ahead of schedule. Various global environmental product regulations such as REACH and China REACH, California Green Chemistry (Safer Consumer Products), Washington Children's Safe Product Act (CSPA), waste electrical and electronic equipment (WEEE), and conflict minerals are discussed. The following new regulations on sustainability initiatives and requirements, like the Corporate Sustainability Due Diligence Directive (CSDDD), Corporate Sustainability Reporting Directive (CSRD), and US Security Exchange Commissions, among others, and their impacts on multinational companies both within and outside of the EU, are reviewed. Company management systems for product regulation compliance and the importance of staying ahead of regulation by having an emerging issue management program are discussed.

A NEW SET OF RULES

It used to be that the only environmental regulations a company had to be concerned with were those affecting a manufacturing facility's air emissions, waste generation, and wastewater. Governments began to realize that the disposal of products presented significant environmental concerns and soon began developing regulations to address this issue. There has been an exponential growth of environmental regulations that apply to products over recent years. Beginning in Europe with requirements for developing more sustainable packaging and mandatory take-back requirements, these regulations have expanded to all regions of the world.

Having to comply with these new types of regulations requires manufacturers to **develop management systems** to ensure that products being brought to

DOI: 10.1201/9781003441939-4

market comply with the myriad of design, reporting, labeling, and fee requirements throughout the world. Organizations within a company that have not typically had to be concerned with environmental regulations now are faced with new challenges. For example, R&D groups have to develop processes to ensure that banned or restricted materials are not in their new products. Sales and marketing units must ensure that labeling, registration, and fees are paid to governments where their products are sold. Further, systems have to be set up to facilitate the recovery and recycling of such things as electrical products, packaging, sharps, and unused medications.

Product-based sustainability regulations are becoming drivers for new product design and are causing **product development teams to anticipate and monitor these requirements**. As a case study, consider having to comply with new requirements that restrict the limit of certain toxic metals, flame retardants, and phthalates in electronics. The European Union Restriction of the Use of Certain Hazardous Substances in Electrical and Electronic Equipment Regulations (the "RoHS Regulations") has expanded into other categories beyond consumer electronics such as medical equipment. Addressing these requirements affects product design criteria.

Knowing that regulatory deadlines are approaching, design groups need to start discussions with their suppliers to begin the testing and validation of new RoHS compliant parts. To prevent barriers of the sale of your product, designs have to be changed years prior to the compliance date. **Companies that can anticipate these regulations and make changes more quickly than the competition can make gains in the marketplace** by appealing to purchasers because they can certify compliance ahead of schedule.

PACKAGING REGULATIONS

Another area of regulation that has complicated product sales is packaging regulations. It is getting increasingly difficult to ensure that product packaging is compliant in all regions of the world. As more companies move toward global brands, a single package is being used for all markets; therefore, the design must incorporate a multitude of regulatory requirements. Consider that there are environmental packaging regulations of some type in all regions of the world, and new regulations are being added on a regular basis.

GLOBAL SUSTAINABLE PACKAGING REGULATORY DEVELOPMENT

- *Plastic packaging*: 83% of the world's sustainable packaging legal measures focus on plastics. The EU and Asia have the highest amount of plastic regulations, with France and India at the top of the list.
- *Beverage packaging*: The EU and North America have 50–60% of their regulatory measures targeting beverages.

- *Food packaging*: Latin America and the Middle East prioritise food packaging.
- *Primary packaging*: Approximately 90% of the world's regulations tackle primary packaging alone or other packaging types.
- *Secondary/tertiary packaging*: Regulations for this type of packaging are more prevalent in Asia, with China, India, Vietnam, and the Philippines at the top of the list.

(*McKinsey 2022*)

Let us consider some of the requirements necessary for a package to be sold globally. As an illustration, let's evaluate the standards of one packaging regulation, the EU Packaging Directive, there's a lot to be concerned with to bring a package on market in the EU!

The EU Packaging and Packaging Waste Directive (*94/62/EC*) requires:

- *Source reduction*: Companies must demonstrate that they have reduced their packaging as much as possible and then identify the critical area (such as product protection, safety, consumer, and acceptance) which prevents further reduction in weight or volume of a packaging component. In 2022, the European Commission presented the revision of *94/62/EC* introducing new regulations to help the EU meet its targets in sustainability. The revision requires all unnecessary packaging to be forbidden.
- *Recovery standards*: Packaging components must be recoverable by at least one of three recovery routes (energy, organic, or material recovery) and must meet certain requirements specific to that recovery route. The terms of the proposal orders packaging producers to have an extended producer responsibility (EPR) and allow for collaborations with organizations that will help to carry out the producer's responsibility.
- *Reuse*: The revision of the Packaging and Packaging Waste Directive *94/62/ EC,* requires all packaging to be reusable or recyclable by 2030 and meet the 2030 and 2040 minimum recycled content targets of plastics. In addition, the packaging items like tea or coffee bags and sticky labels attached to produce must be compostable.
- *Heavy metals content*: Sets a concentration limit for lead, cadmium, mercury, and hexavalent chromium in packaging where the total concentration levels of these combined must not exceed 100 mg/kg.
- *Reduction of hazardous substances in packaging*: Substances classified as noxious (e.g., zinc) must be minimized if they could be released in emissions, ash, or leachate when packaging is land filled or burned.
- *Labelling, marking, and information requirements*: Packaging must contain labels stating information on its composition.

(*Ragonnaud 2023*)

In addition to these requirements, **fees must be paid** based on the type of packaging put on the market. This affects the package design because there are higher fees for packaging that are not easily recyclable. So, to minimize fees, you want to use the least costly (most recyclable) materials in your design. Regulations covering each of these requirements must be developed and put into law for every member state of the European Union, making it more complex to comply for companies placing products on the market in the EU (EUR-Lex 2014).

Similar to the European Union requirement, the government of Korea's Act on the Promotion of Saving and Recycling of Resources and the Resources Recycling Act also has set stringent regulations adopting two schemes regulating packaging materials and controlling packaging waste.

SOUTH KOREA PACKAGING REQUIREMENTS

- Banthe use of PVC materials and colored PET bottles.
- Manufacturers and/or importers are required to review the reliability of their packaging materials and rate them on the following categories: "best to recycle," "good to recycle," "normal to recycle," and "difficult to recycle." Companies that fail to do so are fined and penalized.
- Banthe use of re-packaging, wrapping a final product with film or sheet of synthetic resin.
- Packaging space and layering requirements and regulations on a variety of different products.

Having to conform to **different regulations in many countries makes it very difficult to design a globally compliant package**, and this is just one environmental product regulatory requirement! One of things that my team does at Estée Lauder Companies is to look at global packaging legislation, both existing and pending. We then look for global trends and make recommendations on setting goals or guidelines for new package development to stay ahead of these regulatory requirements. The additional benefit of staying ahead of legislation is that our customers look very favorably at these improvements, like more recyclable packaging and increased PCR, and it helps build brand loyalty.

When you view Figure 3.1 developed by the consulting firm Kearney, consider all the possible steps of a product's life cycle and their potential impacts on the environment. Global regulations are addressing each step of a companies' production to reduce environmental impact, and we are seeing more government regulations on the supply chain and end-of-life of a product and its packaging.

RESTRICTION ON THE USE OF CHEMICALS AND NOTIFICATIONS

One of the most significant chemical regulations that has brought substantial changes to the sale of products in Europe and throughout the world is the **European REACH** regulation. The acronym REACH stands for Registration, Evaluation, and Authorization of Chemicals. The regulation has been in effect since 2007; it requires that all products and chemicals imported and manufactured in EU member

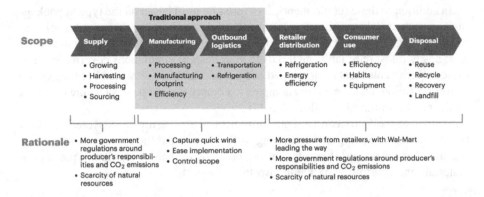

FIGURE 3.1 A product life-cycle approach to sustainability. (Courtesy of A.T. Kearney, Chicago, Illinois, USA.)

states greater than 1 ton per year be registered with the European Chemicals Agency (ECHA). Its aim is to place the responsibility on chemical manufacturers and importers to ensure chemicals are properly tested and are being used in a way that is protective of human health and the environment. The regulation requires more data on the hazards of chemicals developed, and it restricts or bans "substances of very high concern" (SVHC) (*Understanding REACH* 2023). The ECHA continuously updates the REACH candidate list, as of 2023 235 SVHCs have been published (*ECHA adds two hazardous* 2023). This has been a gamechanger for the use of chemicals in products and manufacturing processes and its impact is being seen throughout the world because a lot of companies develop product for sale globally. Besides requiring registrations and much more information on the toxicity of chemicals, companies are facing the prospect of having to find other materials for their products and production processes if they use any chemicals on the SVHC list. The REACH program has recently expanded regulations to include nanomaterials. In 2020, companies were required to disclose information on nanomaterials and their possible risks (*Nanomaterials* 2023). Following the example set in the EU, other countries are adopting REACH-like regulations.

For instance, China and Korea REACH have mimicked certain elements of the European REACH requirements. These regulations are similar to the EU REACH but have their own country requirements. The Measures for Environmental Administration of New Chemical Substances (China MEP, Order 12), also known as **China REACH**, applies to new chemical substances regardless of the quantity. Any new chemical that is not on the Existing Chemical Substances Produced or Imported in China (IECSC) list (about 47,000 substances) must meet certain notification and testing requirements. Some of the requirements of this regulation can be seen in Figure 3.2, where companies who produce higher amounts of chemicals will need to comply with more stringent government regulations.

These REACH-like regulations are becoming more prevalent and are influencing regulatory developments in other regions besides Asia Pacific. A case in point is the California Green Chemistry Initiative, which looked to REACH for inspiration.

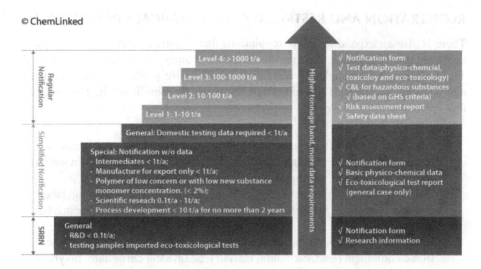

© ChemLinked

FIGURE 3.2 Data requirements for China REACH. (Courtesy of ChemLinked, Hangzhou, Zhejiang, China.)

There are many other regulations that restrict the use of chemicals besides REACH. Several of them focus on electronics and electrical equipment because of the hazardous materials employed in their design. Examples include the European Union RoSH, China, Japan, and Korea RoSH requirements. These regulations put very tight limits on the amount of certain substances that cause harm to human health and the environment if improperly disposed of. As an example, the EU Directive 2002/95 known as RoHS (restriction of hazardous substances) bans the placement of new electrical and electronic equipment containing greater-than-specified levels of lead, cadmium, mercury, hexavalent chromium, polybrominated biphenyl (PBB), and polybrominated diphenyl ether (PBDE) flame retardants, and phthalates on the EU market. The limits range from 0.01% to 0.1% depending on the compound (ROHS Guide 2023).

RoHS restricts the levels for the following materials:

RoHS List of Restricted Substances

Substance Name	Limit (%)
Lead	0.1
Mercury	0.1
Cadmium	0.01
Hexavalent chromium	0.1ch
Polybrominated biphenyls (PBB)	0.1
Polybrominated diphenyl ethers (PBDE)	0.1
Bis(2-ethylhexyl) phthalate (DEHP)	0.1
Butyl benzyl phthalate (BBP)	0.1
Dibutyl phthalate (DBP)	0.1
Diisobutyl phthalate (DIBP)	0.1

REGISTRATION AND RESTRICTION OF CHEMICALS IN PRODUCTS

There is also an expanding list of regulations that require products containing certain hazardous chemicals to be registered in the country where the product is sold.
The following are examples of regulations that require products to be registered,
restricted, or that ban the use of certain chemicals or require labeling of products
that contain specific chemicals.

- *EU REACH*: In effect since 2007; affects all products and chemicals
  imported into and manufactured in EU member states; requirements
  include registration, communication, and restrictions of SVHC.
- *China REACH*: Similar to EU REACH; requires registration, testing, labeling, and data sheets for new chemicals before they can be put on the market
  in China.
- *EU RoHS*: Restricts ten hazardous chemicals used in electrical and electronic equipment (lead, cadmium, mercury, hexavalent chromium, polybrominated biphenyl [PBB] and polybrominated diphenyl ether [PBDE] flame
  retardants), and four phthalates have been in place since 2006.
- *China RoHS*: Restricts the use of the same EU RoHS chemicals not
  including the phthalates; effective since 2007 As of recently in 2022,
  the Chinese Government plans to add the following four new phthalates
  esters, DEHP, BBP, DBP, and DIBP to the restricted list similar to the
  EU RoHS (Wen 2023). The covered products include medical equipment, measuring instruments, radar, communications transmission, and
  switch equipment, and manufacturing equipment for electronic products.
  Requires products to have a label called the Environmental Protection
  Use Period (EPUP) that indicates the number of years it is expected to
  contain hazardous materials without causing environmental contamination (Chemsafetypro 2016).
- *Health Canada*: Focuses on chemicals of concern including lead, mercury, BPA, and phthalates. Canada was the first government in the world
  to restrict BPA in products. They maintain a list of "Chemical Substances
  of Interest," which may lead to the restriction or banning of other materials
  (Health Canada 2023).
- *California Green Chemistry (Safer Consumer Products)*: The California
  Green Chemistry Initiative was signed into law September 2008 and establishes a framework for regulating toxic substances based upon "life cycle
  thinking and green chemistry principles" (Kleen 2012). Formal regulations
  took effect October 1, 2013, with the goals of reducing toxic chemicals in
  consumer products, creating new business opportunities in the emerging
  safer consumer products industry, and helping consumers and businesses
  identify what is in the products they buy for their families and customers.
  In 2022, California's Green Chemistry Program implemented new regulations authorizing the department to rely on publicly available studies of
  alternatives to concerning chemicals when dealing with consumer products
  along with the manufacturing analysis of cautious options. Prior regulations

required manufacturers to seek safer alternatives to harmful chemical ingredients in widely used products (*California Bill Would Change* 2022).

- *Washington Children's Safe Product Act*: Passed in 2008, this requires manufacturers of children's products sold in Washington to report if their product contains a "chemical of high concern to children." The CSPA also limits the amount of lead, cadmium, phthalates, and flame retardants allowed in children's products (SWDE 2023). In response to this regulation, several states have enacted similar laws.
  - *Vermont*: Chemical Disclosure Program for Children's Products
  - *Maine*: Safer Chemicals in Children's Products
  - *Oregon*: The Toxic-Free Kids Act
  - *New York*: Toxic Chemicals in Children's Products
  - *Minnesota*: Toy Safety Ac
- *California Proposition 65*: Proposition 65, the Safe Drinking Water and Toxic Enforcement Act of 1986, was enacted as a ballot initiative in November 1986. The aim of the Proposition is to protect California citizens and the state's drinking water sources from chemicals known to cause cancer, birth defects, or other reproductive harm and to inform citizens about exposures to such chemicals. The rule requires notification and labels so that "no person in the course of doing business shall knowingly and intentionally expose any individual to a chemical known to the state (California) to cause cancer or reproductive toxicity without first giving a clear and reasonable warning." Failure to comply with Proposition 65 will result in costly consequences for business. The California Attorney General reports there were "166 consent judgments resulting in $23.65 million in payments (including attorney fees) and 650 out-of-court settlements that resulted in an additional $26.69 million in payments to private enforcers of Proposition 65" in 2021 (*What are the Consequences* 2022).
- *Interstate Mercury Education & Reduction Clearinghouse (IMERC)*: In 2001, the United States Northeast Waste Management Officials' Association (NEWMOA) launched the IMERC. Anyone who offers to sell, sells, or distributes a mercury-added product in a state covered by this rule is required to complete a notification form and submit it to IMERC. This information informs the public on which products have mercury so that its use may be minimized (NEWMOA 2023).
- *Globally Harmonized System of Classification and Labeling of Chemicals (GHS)*: "The GHS is a system for standardizing and harmonizing the classification and labeling of chemicals." It requires governments around the world to conform their hazard classification and communication rules to a common system, regardless of the country. This initiative will require changes in the way hazards of chemicals are defined and communicated. Consistency will be achieved globally in determining health, physical, and environmental hazards as well as communicating hazards through icons and chemical Safety Data Sheets. For example, countries have different definitions and icons for "flammable." With GHS, the icons and definitions would be the same regardless of the country you operate in (OSHA 2023).

There are many more regulations aside from these that have some form of labeling, restriction, or reporting on the toxic materials used in products. The intent here is not to have an exhaustive list, but to generally inform on the growing amount of regulations that manufacturers must be cognizant of to insure a compliant product on the global market.

**EXAMPLES OF GLOBAL PRODUCT
STEWARDSHIP REGULATIONS**

- *Packaging regulations*: EU Packaging Directive and various others require size reduction, reduction of toxic metals, and take-back.
- *Chemical use restrictions & notifications*: REACH, RoHS, chemical specific bans, e.g., Health Canada BPA ban, IMERC mercury use notification, CA Proposition 65 notifications, and GHS.
- *Extended producer responsibility*: WEEE, Battery, South Africa EPR, etc. require producers to take back products at their end of life.

EXTENDED PRODUCER RESPONSIBILITY

More governments are mandating that manufacturers take responsibility for their products at the end of their useful life and a category of regulations called EPR have been developed. One of the first and most comprehensive regulations of this type is the **European Union Directive** 2012/19/EU known as *waste electrical and electronic equipment* **(WEEE)**. First enacted in 2003 (2002/96/EC), WEEE requires manufacturers to take responsibility for the recycling of electrical equipment at its end of life. The equipment covered by the regulation includes temperature exchange equipment, monitors/equipment with large screens, lamp bulbs, large equipment, small equipment, and small IT/computer/communication equipment. Producers of electrical and electronic equipment are required to set up and pay for collection points where WEEE can be brought to be recycled, and are responsible for the costs of collection, treatment, recovery, and disposal. Equipment is required to be designed to be recycled and must include guidelines for how to recycle the equipment and be labeled to indicate that it is WEEE (*Waste from Electrical* 2023).

The US state of **California** also has a **WEEE** regulation: SB 20. In 2022, the California State Government adopted new laws to expand the e-waste program to include new covered electronic devices, specifically video display devices meeting specific size requirements and battery-embedded products. Batteries contain toxic metals like lead, mercury, and cadmium and if improperly disposed of cause environmental pollution. The new battery EPR program requires the producers to set initiatives for collecting recycling batteries beginning April 2027 (Beveridge & Diamond 2022). A fee to cover recycling is paid for by consumers at the point of purchase. Manufacturers must notify retailers and the California Integrated Waste Management Board (CIWMB) when a device is subject to the recycling fee and provide consumer information on how to recycle the products. Annual reports must be filed with the Board indicating the total amount of hazardous substances in each device, the efforts to reduce hazardous materials, use of recyclable materials, and efforts to design more environmentally friendly products (CalRecycle 2023). California is not the only US state that has electronic equipment take-back requirements; 25 states and the District of Columbia have established electronic take-back and recycling requirement (NCSL 2018).

The **European Union Battery Directive** 2006/66/EC requires the collection and recycling of waste batteries and accumulators at their end of life. In addition, product design must enable batteries to be easily removed to facilitate recycling and include instructions on removing the batteries. Batteries must be labeled with a symbol indicating that they are to be recycled. Product manufacturers must cover the cost of collecting and recycling industrial, automotive, and portable batteries and accumulators, as well as the costs to inform the public of the recycling schemes (EUR-Lex 2014). The EU set in place collection targets that producers are required to follow. For example, the standards for portable batteries are as follows: 45% by 2023, 63% by 2027, and 73% by 2030. The European Parliament has set minimum requirements for reusing cobalt, lead, nickel, and lithium in new batteries (*Making Batteries More* 2023).

On May 5, 2021, **South Africa** passed section 18 of the Natural Environment Management Waste Act to introduce EPR in tackling package waste in the environment. The Department of Forestry, Fisheries, and Environment has identified PET, polyolefins, polystyrene, vinyl, glass, paper, and metals as products needed to be supervised by an appropriate EPR scheme. The EPR policy objectives are to change South Africa's economic activities, urging companies to take action financially and physically on the issue of waste prevention. Producers must meet the strict government requirements on collection and recycling for future years (WWF 2024) **South Africa and Kenya** are among the first countries to set EPR regulations in Africa, helping pave the way for other countries in combating the issue of rising pollution (Langhill 2021).

Numerous EPR regulations have been developed in the United States. As seen in Figure 3.3, requirements have been established at **the state level.** These regulations are cropping up throughout the United States, but this is truly a global phenomenon that is building momentum.

The development of regulations for take-back of products and forcing companies to be responsible for the end of life of their products is becoming more popular with legislators. This policy initiative enables products to be properly managed when they have lost their usefulness and typically at a cost that is borne by manufacturers, which adds to its attractiveness. We should expect to see more of this type of regulation in all markets throughout the world; this is especially true considering the new focus on developing a **circular economy**.

ENVIRONMENTAL SOCIAL GOVERNANCE (ESG) DISCLOSURE REGULATIONS

As highlighted in Chapter 2, businesses worldwide are voluntarily addressing ESG issues to remain competitive in the ever-changing market. Today, consumers, activists, and investors demand transparency from companies regarding their ESG

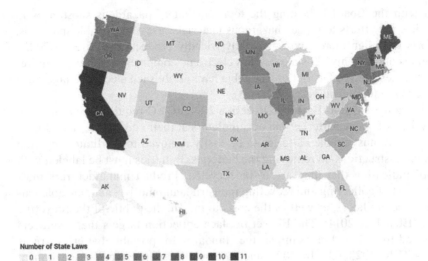

Number of State Laws

0 1 2 3 4 5 6 7 8 9 10 11

This map does not include local laws, nor does it reflect disposal and sale bans on products containing toxic materials, policies that require producers to collect products for recycling, or those that require the purchase of environmentally preferable products. We have not included deposit fee or container deposit laws but please note that, although they are technically not EPR, "bottle bills" do shift responsibility upstream to producers and are highly effective at recovering beverage containers to support recycling markets. Read more at www.productstewardship.us

Map: 2022 Source and Copyright © Product Stewardship Institute, Inc. • Created with Datawrapper

FIGURE 3.3 EPR regulations across America. (Courtesy of Product Stewardship Institute, Boston, Massachusetts, USA.)

commitments and their tangible actions. Additionally, governmental agencies are exerting pressure on businesses to prioritize and assume greater responsibility for ESG initiatives. ESG disclosure has for years been a strictly voluntary initiative; however, there has been a new focus on development of regulations. Below are some examples of laws that demonstrate these global phenomena.

The EU Corporate Sustainability Due Diligence Directive (*CSDDD*)—In February 2022, the European Commission adopted legislation requiring companies to take responsibility for their global environmental and social impacts. Both **EU and non-EU-based companies** are required to establish policies and procedures that demonstrate how they are protecting the environment and human rights.

The EU Corporate Sustainability Reporting Directive (*CSRD*)—Enacted in January 2023, the new legislation requires companies to **disclose information on sustainability issues** and verify this information by third-party organizations. For example, companies must communicate their business models/strategies to sustainability risks and their current actions regarding the 1.5°C global warming target under the Paris Agreement. Other reporting requirements are on the protection of water and marine resources, circular economy, pollution prevention and control, and protection of biodiversity.

The UK Task Force on Climate-Related Financial Disclosures (*TCFD*)—Put in effect in April 2022, the United Kingdom became the first G-20 country to make **mandatory requirements** for companies to report **climate management efforts and risks**. Greg Hands, the Energy and Climate Change Minister stated: "If the UK is to meet our ambitious net-zero commitments by 2050, we need our thriving financial system, including our largest businesses and investors, to put climate change at the heart of their activities and decision making."

The US Security and Exchange Commission (SEC)—In March 2021, the SEC announced the new Enforcement Task Force Focused on Climate and ESG issues to help identify ESG-related misconduct consistent with increased investor reliance on such disclosures. The purpose of this program is to identify financial reporting violations including any failure to address climate-related gaps in material (significant) impacts or any misstatements made for publicly traded businesses.

California Global Warming Solutions Act (AB 32)—The Regulation requires the mandatory reporting of Greenhouse Gas Emissions for companies doing business in California. Newer regulation was adopted in 2023, Climate Corporate Data Accountability Act (SB-253) and the Greenhouse Gases: Climate-Related Financial Risk Act (SB-261). Companies are to report their Scope 1, 2, and 3 greenhouse gas emissions, have them assured by an independent firm, and report on financial risks related to climate change.

Hong Kong's Guidance on Climate Disclosure—Inspired by the TCFD, the Hong Kong government established the legislation in November 2021, to help companies assess and disclose their response to the risks revolving around climate change. The Hong Kong Exchanges and Clearing Limited (HKEX) expects the companies covered by the law to report annually by 2025.

At Estée Lauder Companies (ELC), we are required to report both to the SEC and the EU CSRD on the company's ESG initiatives even though we are headquartered in the United States. In addition to these regulatory initiatives, we perform a Priority Issues Analysis and gather information on sustainability and social issues that are relevant to the business that we put into our annual Social Impact and Sustainability report. This analysis enables us to report key sustainability issues for our business using the voluntary reporting systems: Global Reporting Initiatives (GRI), International Sustainability Standards Boards (ISSB), and the TCFD.

SUPPLY-CHAIN ISSUES

Focus on environmental and social issues deep in the supply chain is a new regulatory initiative. **Manufacturers are being held accountable not just for their own operations, but also that of their suppliers**. It is becoming increasingly important to know what's in your product to very specific details like the farm that raw materials are harvested. This goes beyond the materials used in the product and includes social issues like the working conditions at suppliers, how suppliers' employees are treated, and the living conditions of where raw materials are sourced.

In California SB 657, the **California Transparency in Supply Chains Act** requires companies doing business in California with more than $100 million in gross receipts to report on efforts to **eradicate slavery and human trafficking** from its supply chain. The reporting requirements are very broad and go into areas that governments never before considered. Disclosures of activities to ensure that human trafficking is not occurring in a product's supply chain must be reported on the company's website. The information requires disclosure on how the business:

1. Engages in verification of supply chains to address human trafficking and slavery and uses third parties in that process.
2. Conducts independent, unannounced audits of suppliers to ensure compliance with company standards on trafficking and slavery.
3. Requires direct suppliers to certify that materials incorporated in their products comply with the laws regarding slavery and human trafficking of the country or countries in which they do business.
4. Maintains internal accountability standards for employees and contractors failing to meet company standards on slavery and trafficking.
5. Provides both managerial and non-managerial employees with training on mitigating risks of slavery and trafficking in supply chains

(State of California Department of Justice, 2023)

In the United Kingdom, **the Modern Slavery Act** requires businesses to release annual statements confirming their actions in regard to slavery and human trafficking. The legislation's objective is to support/protect the victims and punish those responsible. A few of the offences listed under the act are as follows:

- "the person holds another person in slavery or servitude and the circumstances are such that the person knows or ought to know that the other person is held in slavery or servitude".
- "A person commits an offence if the person arranges or facilitates the travel of another person with a view to being exploited".

(Tackling Modern Slavery 2023)

The legislation introduces measures in which companies must create accountability standards, conduct evaluations of supplier compliance, and provide training on human trafficking and slavery. The act also includes trafficking reparations directives encouraging companies to compensate victims through their seized assets and set prevention practices to phase out suppliers likely to commit such crimes (EcoVadis 2023).

The European Commission is also taking action in promoting **human rights in the supply chain**. In September 2022, the commission presented a proposal for regulation to prohibit products made with forced labor in the EU (*Commission Moves to Ban* 2022).

This view into a company's supply chain is **adding more scrutiny to every step of bringing a product to market**. Governments are responding to concerns with the conditions of the suppliers used for making products and are holding manufacturers accountable to ensure that their product does not enable horrendous conditions.

Another similar initiative is the use of conflict minerals. If someone would have told me that I would be working on **conflict minerals** as part of our product steward-ship program, I would have thought that had nothing to do with our company. **Where raw materials are sourced from is a growing concern** of procurement and product stewardship teams. Various products contain minerals that come from areas where there is conflict, such as Africa. Cell phones, laptop computers, televisions, medical equipment, and many other products contain these minerals.

The US government through the "conflict minerals" provision of the Dodd-Frank Act requires companies to publicly disclose if they use these materials. Section 15.2(e) (4) of the Dodd-Frank Act defines "conflict mineral" as cassiterite (tin), wol-framite (tungsten), coltan (tantalum), and gold (gold) or their derivatives. Public com-panies are to report on minerals used in their manufactured goods that originated in war-torn Congo or adjoining countries in Africa. "Companies are not encouraged to stop sourcing from this region but are required to show they are working with the appropriate care—what is now known as 'due diligence'—to make sure they are not funding armed groups or human rights abuses" (Global Witness 2017).

Another area that affects the sourcing of products in a significant way is **bio-piracy**, also known as **access and benefit sharing** (ABS) regulations. The **Nagoya Protocol** on Access to Genetic Resources and the Fair and Equitable Sharing of Benefits Arising from their Utilization was an outcome of the United Nations Convention on Biological Diversity. The idea of this regulation is to con-sider "the need to **share costs and benefits between developed and developing countries**." In the past, developing countries have had genetic resources like the use of unique plant-based raw materials used for consumer products, like natural ingre-dients and even pharmaceutical ingredients, removed from their country without permission or even compensation. Approvals and benefits are to occur for the use of these raw materials. Benefits may be monetary or non-monetary such as royalties and the sharing of research results. The Nagoya Protocol creates greater legal cer-tainty and transparency for both providers and users of genetic resources by sharing the benefit of these resources and making it more predictable for the users of these materials. If you are a product manufacturer and are using ingredients such as natu-ral or botanical extracts which are unique to a specific country, you may be subject to these requirements (CBD 2023).

This new focus on the supply chain is a continuation of a trend that requires product manufactures to know the minutest details about how products are brought to market. Doing this can be extremely difficult since parts that go into products come from sev-eral tiers of suppliers, going back to the mine where minerals came from is not readily available information. Consider the many materials that go into one part and that some products have hundreds of parts. Now try to understand where all the materials came from. This is a daunting task; nevertheless, **supply-chain regulations are most likely going to increase** and cover areas that today we cannot even imagine.

COMPANY MANAGEMENT SYSTEMS FOR PRODUCT REGULATIONS

With all of the new regulatory requirements that apply to products, how does a global company ensure compliance? As mentioned earlier, more and more businesses are

developing and marketing products that are sold globally. Having a robust system to
ensure that products being developed are meeting national standards is becoming
increasingly more important. Management systems must be put in place for product
development groups both in-house and for third-party partners, to ensure compliance. As noted above, there are an enormous set of requirements that must be com-
plied with.

One approach to manage this complexity is the use of company developed standards. A good example of this is **Hewlett-Packard's (HP's) General Specification
for the Environment**. To ensure that products meet the global rules, it requires
compliance to this standard in all contracts for design, manufacture, or purchase of
HP brand products. This includes subassemblies, parts, materials, components, batteries, and packaging that are incorporated into HP brand products.

The standard has **121 pages of requirements!** There is a section that addresses
"Substances and materials requirements." HP restricts or prohibits certain materials
in their parts and products. Some examples include like Halogenated flame retardants and polyvinyl chloride (PVC) must not be intentionally added and it is limited
to 0.09% 900 ppm, hexavalent chromium and its compounds in metallic applications
must not be present, mercury and its compounds 0.1% (1000 ppm), PVC in external
case plastic parts must not be present. Mercury must not be intentionally added in
any battery, and it must not contain more than 0.0001% (1 ppm) mercury by weight.
Tributyl tin, triphenyl tin, and tributyl tin oxide must not be used in parts, components,
materials, or products. The use of wood is also addressed. Parts, components, materials, and products must not contain any wood material or wild plant material that was
illegally sourced from its country of origin. Heavy metals must be managed in packaging; the sum concentration of incidental lead, mercury, cadmium, and hexavalent
chromium may not be greater than 0.01% (100 ppm) by weight. Product stewardship
regulations are also addressed. Compliance must be achieved with any applicable regulations and documentation must be maintained. Examples include China RoHS User
Documentation, Korea e-Standby User Instructions for Personal Computers, Product
End of Life Declarations, and many, many more requirements (HP 2022).

The use of a management system such as this HP Standard is a good example
of the type of processes that have to be put in place to ensure developers and parts
suppliers are providing compliant products onto the global marketplace. The fact
that this document is over 100 pages long speaks to the difficulties of addressing
all the requirements out there. One thing is for sure: Companies must develop some
mechanism to track and assist R&D groups and suppliers on how to comply with the
multitude of global requirements.

MANAGING RISK

As we have seen, regulations can have a huge impact on a company's ability to com-
pete in the marketplace. Corporations are always trying to minimize their business
risk and have adopted various methods to mitigate and anticipate adverse situations
that can interrupt or even stop products from being sold. One of the most difficult
situations to be in is when an issue comes out of nowhere, and you're completely
blindsided by it. Even worse is having to defend yourselves in the court of public

opinion because of it. As one of my company's business leaders used to say, "When you're explaining, you're losing."

As one of my company's business leaders used to say; "When you're explaining—you're losing!"

Business managers like to think that if you have good scientific evidence on your side, this is all you need to prove your point of view is correct. However, time after time we have seen that perception trumps scientific evidence. Once an issue gets out into the public in a widely distributed report, whether it's based on scientific evidence or not, it becomes very difficult to dispute after the fact. A good way of looking at risk associated with a product is to understand that risk = hazard + outrage (Blake 1995).

RISK = HAZARD + OUTRAGE

The leading factor that seems to bring public distrust of any industry is **a lack of transparency**. The public is leery of firms that don't appear forthright with information and are not telling them all they know about an issue. A good example of an issue that has generated a lot of angst for some companies is genetically modified organisms (GMO). When the use of GMO crops first came to the public eye, there was outrage. Cries of Franken-food and environmental-group protests were prevalent. However, GMO crops have the possibility to address food shortages by being resistant to bad weather, insects, or viruses (HGP 2008). Regardless of where you stand on the issue, it makes you wonder if the companies that developed these GMO crops had an emerging issues management program—did they see the outrage coming? Could they have been more transparent? Would it have been possible to have reached out to the opposition groups and share data? I don't have the answers to this, but a robust emerging issues process would have given this issue a better chance of not being the firestorm that it was and continues to be.

Factors that Decrease Risk	Factors that Increase Risk
Voluntary	Imposed
Control	Lack of control
Fair	Unfair
Ordinary	Memorable
Not dreaded	Dreaded
Natural	Technological, artificial
Certain	Uncertain
Familiar	Unfamiliar
Morally acceptable	Morally unacceptable
Trustworthy source	Untrustworthy source

Source: Blake (1995).

CORPORATE REPUTATION

During my career at Johnson & Johnson we had an exceptional CEO named Ralph Larsen. At one of our executive leadership meetings he reinforced the importance of our corporate reputation. Our name was not a trademark but a **trust mark**, and reputation is like a bank account, you make deposits and there are withdrawals, therefore we better make sure that we have a lot of deposits in our account to ensure that we can withstand all assaults that will come. This sentiment aligns well with the principles espoused in the book entitled, *The 18 Immutable Laws of Corporate Reputation*. Similar to what Mr. Larsen said, it speaks to reputation being like a life preserver and is a "tailwind when you have opportunity." One of the laws is to "learn to play to many audiences," and one of those audiences are environmental groups. An example of doing this right was when fast food companies were pressured on environmental issues and nutrition. Instead of fostering an adversarial relationship, they worked with the pressure groups and developed practices that treated the farm animals that were used for food in their stores more ethically. This was a smart move and garnered positive comments from the groups that were putting pressure on them. I have personally been involved in working with various environmental groups and when you sit down and listen, you find out that there are common issues that usually can be worked through.

A good concept relevant to product concerns addressed in the above-mentioned book is to get in front of issues. When a controversial issue is raised against your company, it's usually a good idea to lead and proactively address the issue and speak publicly about what you're doing about it. If you do not take this tack, you may find your company on the wrong end of a boycott or taking a major reputational hit. Another "law" is to "control the Internet before it controls you." This concept connects to the proactive approach; we see many initial studies or campaigns initiated on the Internet. It's a wise move to monitor the blogs and websites of key environmental groups and influencers. Being vigilant and taking every perceived threat seriously is a good way to stay **ahead of the curve** and manage emerging issues that can affect a company's reputation (Alsop 2004).

EMERGING ISSUES MANAGEMENT

It is possible to get signals on emerging issues that can result in business risk and appropriately influence their development. **An issue takes a certain path before it becomes a regulation or a public relations nightmare**. There are opportunities to address issues when they are in the "anticipatory" stage. Monitoring the warning signs coming from NGO reports, blogs, web postings, and research reports are imperative to getting a read on any developing business risk.

Emerging issues must be put through a filter that considers

1. Potential business impact (both financial and reputational).
2. The time it will take to become a "crisis" and hit the public in the form of a news report or a regulation.

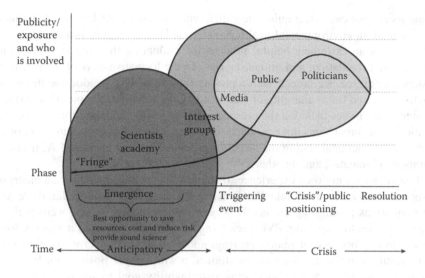

FIGURE 3.4 The general life cycle of an issue.

Resources, such as the commissioning of teams of experts to study and recommend actions to management, should be deployed on the issues that have the greatest potential to (negatively) impact the business. **Actions are most effective if they occur in the early stages** of the "anticipatory" stage; see the chart in Figure 3.4.

Some of the outcomes of an effective emerging issues process could include:

- Commissioning research on topics to develop science.
- Papers developed and posted in peer-reviewed scientific journals to develop the science on the new issue.
- Presentations of data and perspectives at scientific conferences.
- White papers and position papers that discuss the issue and what the organizations' understanding and positions are on the issue.
- Communications in the form of meetings with NGOs or government agencies to share knowledge and points of view on issues.
- Development of company guidelines and standards.

An emerging issues process must have a strong connection to the company's government affairs group since they see what legislatures are considering first. Granted, this is much further down the path toward an issue becoming law, but it is critical to have sound scientific principles, front of mind, when regulations are being formed. Working with legislatures in the earliest stages possible is also necessary to prevent unnecessary regulatory burdens that do not add value.

EXAMPLES OF EMERGING ISSUES

What type of issues are being monitored by companies today that can have significant business risk in the future? It depends a lot on the type of industry, but there are some issues that can affect quite a few different companies. At Johnson & Johnson, we initiated an environmental health & safety (EHS) **emerging issues process** in 1998; it was an extremely helpful process for evaluating the landscape, influencing potential regulation, and mitigating risk. In the formative days of the Emerging Issues Committee, we tagged such topics as the use of PVC, endocrine disrupting compounds, and the availability of freshwater. Over time, other topics that surfaced as significant issues included the presence of very small concentrations of pharmaceuticals and other consumer products in water, nanotechnology, bio-monitoring, and one-off chemicals being the target of pressure, such as bisphenol A, triclosan, parabens, phthalates, among others.

I can draw on my own experience to share a few examples of how an emerging issues process works. Let's take **PVC**—when it first arrived on our radar, there were no regulations prohibiting its use. However, environmental groups were petitioning companies to stop using PVC packaging because they claimed it was the worst plastic to use because of manufacturing, and end-of-life concerns. Work on this issue resulted in the development of studies, a white paper position on PVC, and eventually, adoption of a public-facing sustainability goal to significantly reduce and, in some cases, completely eliminate PVC from company packaging (we eliminated over 3,000 tons of PVC from Johnson & Johnson packaging). This approach paid dividends because our customers were asking for products and packaging free of PVC and we were ahead on this issue because of our emerging issues process. At Estée Lauder Companies we picked up on single use plastic packaging being a growing concern as part of what I call the **war on plastic.** Consumers, NGOs, and investors were all growing concerned with plastic use, and it is making its way into the ocean. Seeing this issue as having great acceleration and a big potential business impact we set an external facing goal to have 50% of our plastic packaging to be absent of virgin petroleum plastic by 2030.

Nanotechnology is an example of an emerging issue. Governments are trying to determine the best way to address the use of compounds that are very common, but now can be milled down to nanometer size (nano size = 1–100 nanometers in dimension). Nanomaterials can be found in everyday products, stimulating much innovation in the world of science. However, questions have been raised as to whether smaller sized materials will react differently in the environment or cause significant human health issues compared to larger-sized particles. If your company is using nanomaterials, it would be in your best interest to understand if there are increased impact and risk of their use in your product (*Control of Nanoscale Materials* 2023).

SOME CURRENT EMERGING ISSUES

Biomonitoring: Finding of trace chemicals in human blood and body fluids.

Biodiversity: Impact on the variety of life or living beings on Earth.

Chemicals of concern: Several individual chemicals being pressured by NGOs as harmful to human health & the environment, for example, BPA, triclosan, phthalates, DEHP, etc.

> *Chemical mixtures*: Minute concentrations of chemicals in the environment and their collective impact.
>
> *Climate change*: Impacts of climate change on raw material supplies and other potential business disruption.
>
> Endocrine disrupting chemicals (EDCs): Fate of EDCs in the environment.
>
> *Human rights*: These rights inherent to all human beings, such as freedom of slavery and torture and child labor.
>
> *Nanotechnology*: 1–100 nanometers in dimension materials impact on human health and the environment.
>
> *PPCPs*: Trace amount of personal care products and pharmaceutical ingredients found in the environment.
>
> *Water scarcity*: Availability of safe and secure water supply.

In addition to nanotechnology, governments and companies worldwide are addressing the concerns surrounding per- and polyfluoroalkyl substances (**PFAS**). PFAS chemicals nicknamed "forever chemicals" are fluorinated carbon chains attached to functional groups creating much stability and strength. The substances are difficult to break down in the environment, have the power to bioaccumulate, and enter the water supply. Due to these chemical properties, PFAS are used in many everyday products for resistant coatings like nonstock cookware, food packaging, outdoor gear, stain resistant furniture, and many household items (*What are PFAS* 2022). Studies demonstrate possible health risks among the population which include reproductive effects, developmental effects, increased risks of certain chemicals, interference with the endocrine system, and increased levels of cholesterol (*What are the Health Effects* 2022). Since January 21, 2022, the US EPA has been researching and taking action on the problems of PFAS and have set numerous regulations on PFAS in water (*Key EPA Actions* 2023). In 2023, the ECHA set proposals to limit PFAS emissions in the environment (ECHA publishes PFAS restriction 2023). This is yet another example of getting signals about an issue and putting processes in place to address and manage an emerging issue before it damages your business.

DISCUSSING EMERGING ISSUES PUBLICLY

As mentioned above, an important aspect to preventing outrage on emerging issues is dialog. One way to allay focus on issues of concern is for companies to discuss dilemmas publicly. By doing so, it lets stakeholders know where they stand and that they are paying attention to issues they are concerned with. Many companies are now implementing actions against PFAS, implementing policies to fully ban the chemicals from their products. Among the fast-food industry, McDonald's, Starbucks, and Chipotle, among others, are phasing out PFAS from food packaging. Retailers like Home Depot, Target, Staples as well as others are introducing new policies in restricting PFAS from their stores and even setting bans (Toxic-Free Future 2023).

Another good example of companies taking public action on upcoming environmental problems can be found on Baxter's website. Being a medical products manufacturer, Baxter had received some adverse publicity about the use of PVC in some of their products. The use of PVC is discussed on their website through a Materials Use Position Statement. This statement addresses exactly how the company sees the use of PVC in medical products and what they are doing to address customer concerns regarding this material.

BAXTER ADDRESSES PVC

PVC has a long history of use in a variety of medical products, such as contact lenses, intravenous bags, oxygen tents and catheters. These products have undergone strict regulatory review by many government and independent health agencies throughout the world, including the U.S. Food and Drug Administration. The safety of these materials has been confirmed by more than 40 years of use with approximately eight billion patient days of acute and chronic exposure without report of significant adverse effects. Environmental questions relating to the incineration of PVC are being addressed with modern pollution control technologies that can reduce, for instance, dioxin emissions up to 99.9%.

While PVC is a widely used material that can consistently meet the rigorous requirements for medical products, **it may not be appropriate for all clinical applications**. To meet the **preferences of some customers** and address drug compatibility issues in specific clinical applications, **Baxter offers a portfolio** of intravenous medications, parenteral nutrition solutions, injectable drugs, biopharmaceuticals, IV sets and access devices and other products that use or are contained in non-PVC materials or non-DEHP [di-(2-ethylhexyl)phthalate], a common component of PVC, materials.

(Baxter 2023)

This is a very good example of publicly stating a position on a topic of concern to stakeholders and demonstrating that consideration on the topic was made and non-PVC alternatives are being pursued. Discussing emerging issues publicly that are of concern to your customer is a great way to address these challenges. Being transparent builds trust, and that's a really good thing to develop loyalty with your customers.

CONCLUSION

Environmental regulatory requirements for products are increasing in all regions of the world. Governments are realizing that they can address many environmental and human health issues by focusing on products. Many have started in Europe and have inspired other governments to write similar regulatory requirements, such as REACH, RoHS, WEEE, and packaging regulations. Besides making regulatory compliance more complex, these expanding requirements are becoming a significant driver of greener product development. Having to consider requirements that affect product design of packaging, toxic material use in products, and producer responsibility requirements for product end-of-life management all result in forcing greener product design. Some companies have enacted robust management systems to ensure that product developers and suppliers are aware of and anticipating regulatory requirements in their design. As these requirements become more important for products that are marketed globally, manufacturers are trying to track and influence the development of issues that can affect products' marketability through emerging issues processes. It is evident that the increase of environmental product legislation will make tracking and complying with these new requirements a critical part of bringing greener products to market.

REFERENCES

Alsop, Ronald J. 2004. *The 18 Immutable Laws of Corporate Reputation*. Free Press, New York, p. 17, 45, 75, 134.

Blake, Elinor R. 1995. Understanding Outrage: How Scientists Can Help Bridge the Risk Perception Gap. *Environmental Health Perspectives 103*, 123–125.

World Wildlife Fund (WWF).| Development from a voluntary to a mandatory EPR scheme for packaging. *Alliance*. https://wwfafrica.awsassets.panda.org/downloads/epr_synthesis_report.pdf?34924/Extended-Producer-Responsibility-for-plastic-packaging-in-South-Africa (Accessed April 20, 2024)

California Bill Would Change Requirements for Alternative Assessment under Green Chemistry Program. 2022. Packaging Law. https://www.packaginglaw.com/news/california-bill-would-change-requirements-alternative-assessment-under-green-chemistry-program (Accessed June 12, 2023).

California Passes Two New Laws to Overhaul State's Battery Extended Producer Responsibility Program and Broadly Expand State's E-Waste Program. 2022. Beveridge & Diamond. https://www.bdlaw.com/publications/california-passes-two-new-laws-to-overhaul-states-battery-extended-producer-responsibility-program-and-broadly-expand-states-e-waste-program/ (Accessed June 8, 2023).

The California Transparency in Supply Chains Act. 2023. State of California Department of Justice. https://oag.ca.gov/SB657 (Accessed June 8, 2023).

Chemical Substances. 2023. Health Canada. https://www.canada.ca/en/health-canada/services/chemical-substances.html (Accessed July 18, 2023).

Children's Safe Product Act. 2023. State of Washington Department of Ecology (SWDE). http://www.ecy.wa.gov/programs/hwtr/RTT/cspa/index.html (Accessed July 19, 2023).

China RoHS2. 2016. Chemsafetypro. http://www.chemsafetypro.com/Topics/Restriction/China_RoHS_2_vs_EU_RoHS_2.html (Accessed December 17, 2016).

Commission moves to ban products made with forced labour on the EU market. 2022. European Commission. https://ec.europa.eu/commission/presscorner/detail/en/ip_22_5415?utm_campaign=greenbuzz&utm_medium=email&utm_source=newsletter&mkt_tok=MjExLU5KWS0xNjUAAAGMTqxVEx33m2YOvjf6JWBwXzvBgS8mxOnLcllv3ntSFVOflNco24PQUWlEDxoWWDfGUVOW-V6aSs88U7ffFeMvxOMNLfqwv7myBwVanqMFoPolKRs (Accessed June 26, 2023).

Control of Nanoscale Materials under the Toxic Substances. 2023. US Environmental Protection Agency. https://www.epa.gov/reviewing-new-chemicals-under-toxic-substances-control-act-tsca/control-nanoscale-materials-under (Accessed June 29, 2023).

Convention on Biological Diversity. 2023. CBD. https://www.cbd.int/abs/ (Accessed July 18, 2023).

Corporate Sustainability Due Diligence. 2023. European Commission. https://commission.europa.eu/business-economy-euro/doing-business-eu/corporate-sustainability-due-diligence_en (Accessed June 20, 2023).

Corporate Sustainability Reporting. 2023. European Commission. https://finance.ec.europa.eu/capital-markets-union-and-financial-markets/company-reporting-and-auditing/company-reporting/corporate-sustainability-reporting_en (Accessed June 26, 2023).

Disposal of Spent Batteries and Accumulators. 2014. EUR-Lex. http://eur-lex.europa.eu/legal-content/EN/TXT/?uri=URISERV:l21202 (Accessed December 30, 2016).

ECHA adds two hazardous chemicals to Candidate List. 2023. European Chemicals Agency. https://echa.europa.eu/en/-/echa-adds-two-hazardous-chemicals-to-candidate-list (Accessed July 17, 2023).

ECHA publishes PFAS restriction proposal. 2023. European Chemicals Agency. https://echa.europa.eu/en/-/echa-publishes-pfas-restriction-proposal#:~:text=ECHA%20publishes%20PFAS%20restriction%20proposal&text=Helsinki%2C%207%20February%202023%20%E2%80%93%20The,and%20processes%20safer%20for%20people. (Accessed July 18, 2023).

Electronics Take Back Coalition. 2016. *State Legislation.* http://www.electronicstakeback.com/promote-good-laws/state-legislation/ (Accessed December 11, 2016).

Electronic Waste Recycling. 2018. National Conference of State Legislatures. https://www.ncsl.org/environment-and-natural-resources/electronic-waste-recycling (Accessed June 22, 2023).

Electronic Waste Recycling Act of 2003. 2023. CalRecycle. https://calrecycle.ca.gov/electronics/statutes/ (Accessed January 3, 2017).

Enforcement Task Force Focused on Climate and ESG Issues. 2023. U.S. Securities and Exchange Commission. https://www.sec.gov/securities-topics/enforcement-task-force-focused-climate-esg-issues (Accessed June 26, 2023).

ESG Disclosure Regulations Are Strengthening in Asia Pacific. 2022. Schneider Electric. https://newsdirect.com/news/esg-disclosure-regulations-are-strengthening-in-asia-pacific-781581983 (Accessed June 21, 2023).

Foley & Lardner. 2011. *New California Employment Regulations: Is the "Reprieve" Over?* (Accessed January 3, 2017).

HP Standard 011 General Specification for the Environment. 2022. Hewlett-Packard Development Company, L.P. https://h20195.www2.hp.com/v2/getpdf.aspx/c04932490.pdf (Accessed July 18, 2022).

Human Genome Project (HGP). 2008. Genetically Modified Foods and Organisms. November 05. http://theliteratesims.net/eng1bM/Readings/gmfoodsandorganisms.pdf (Accessed December 28, 2016).

Key EPA Actions to Address PFAS. 2023. US Environmental Protection Agency. https://www.epa.gov/pfas/key-epa-actions-address-pfas (Accessed June 12, 2023).

Kleen, C. 2012. RoHS & REACH. *American Galvanizers Association.* https://galvanizeit.org/knowledgebase/article/rohs-reach#:~:text=It%20is%20known%20as%20the,thinking%20and%20green%20chemistry%20principles (Accessed July 18, 2023).

Langhill, R. 2021. EPR in Africa – what to expect in the next few years. *Lorax EPI.* https://www.loraxcompliance.com/blog/env/2021/07/07/EPR_in_Africa_-_what_to_expect_in_the_next_few_years.html (Accessed June 23, 2023).

Making Batteries More Sustainable, More Durable and Better-Performing. 2023. European Parliament. https://www.europarl.europa.eu/news/en/press-room/20230609IPR96210/making-batteries-more-sustainable-more-durable-and-better-performing (Accessed June 29, 2023).

Medical Materials Use Position Statement. 2023. Baxter. https://www.baxter.com/policies-positions/medical-materials-use (Accessed July 18, 2023).

Nanomaterials. 2023. European Chemicals Agency. https://echa.europa.eu/en/regulations/nanomaterials (Accessed June 29, 2023).

NEWMOA. 2023. *Interstate Mercury Education & Reduction Clearinghouse (IMERC).* http://www.newmoa.org/prevention/mercury/imerc.cfm (Accessed July 18, 2023).

Occupational Safety and Health Administration (OSHA). 2023. *Globally Harmonized System of Classification & Labeling.* https://www.osha.gov/dsg/hazcom/global.html (Accessed July 18, 2023).

Packaging and Packaging Waste. 2014. EUR-Lex. http://eur-lex.europa.eu/legal-content/EN/ TXT/?uri=URISERV:l21207 (Accessed January 3, 2017).

Ragonnaud, G. 2023. Revision of the Packaging and Packaging Waste Directive. European Parliament. https://www.europarl.europa.eu/RegData/etudes/BRIE/2023/745707/EPRS_ BRI(2023)745707_EN.pdf (Accessed June 13, 2023).

REACH. 2023. European Commission. http://ec.europa.eu/environment/chemicals/reach/ reach_en.htm (Accessed December 10, 2016).

Retailers committing to phase out PFAS as a class in food packaging and products. 2023. Toxic-Free Future. https://toxicfreefuture.org/mind-the-store/retailers-committing-to-phase-out-pfas-as-a-class-in-food-packaging-and-products/ (Accessed June 12, 2023).

Sustainability in packaging: Global regulatory development across 30 countries. 2022. McKinsey & Company. https://www.mckinsey.com/industries/paper-forest-products-and-packaging/our-insights/sustainability-in-packaging-global-regulatory-development-across-30-countries (Accessed June 8, 2023).

Tackling Modern Slavery in Government Supply Chains-Guidance. 2023. The UK Government. https://www.gov.uk/government/publications/ppn-0223-tackling-modern-slavery-in-government-supply-chains/ppn-0223-tackling-modern-slavery-in-government-supply-chains-guidance-html#section-1—introduction (Accessed July 18, 2023).

Tao, L. 2020. South Korea Bans the Use of PVC Packaging and Colored Plastic Bottles. *ChemLinked.* https://food.chemlinked.com/news/food-news/south-korea-bans-use-pvc-packaging-and-colored-plastic-bottles (Accessed July 16, 2023).

UK Modern Slavery Act. 2023. Ecovadis. https://ecovadis.com/glossary/uk-modern-slavery-act/ (Accessed June 26, 2023).

Understanding REACH. 2023. European Chemicals Agency. https://echa.europa.eu/en/ regulations/reach/understanding-reach (Accessed July 17, 2023).

US Conflict Minerals Law. 2017. Global Witness. https://www.globalwitness.org/en/ campaigns/conflict-minerals/dodd-frank-act-section-1502/ (Accessed July 18, 2023).

Waste from Electrical and Electronic Equipment. 2023. September 6. European Commission http://ec.europa.eu/environment/waste/weee/index_en.htm (Accessed July 18, 2023).

Wen, T. 2023. China Consults on Revision of RoHS Industry Standard. *ChemLinked.* https:// chemical.chemlinked.com/news/chemical-news/china-consults-on-revision-of-rohs-industry-standard (Accessed June 21, 2023).

Welcome to ROHS Guide. 2023. ROHS Guide. https://www.rohsguide.com/ (Accessed July 17, 2023).

What are the Consequences of Omitting a Required Proposition 65 Warning? 2022. Packaging Law. https://www.packaginglaw.com/ask-an-attorney/what-are-consequences-omitting-required-proposition-65-warning#:~:text=In%202021%2C%20 alone%2C%20the%20California,private%20enforcers%20of%20Proposition%2065 (Accessed July 17, 2023).

What are the Health Effects of PFAS? 2022. Agency for Toxic Substances and Disease Registry (ATSDR). https://www.atsdr.cdc.gov/pfas/health-effects/index.html (Accessed June 29, 2023).

What are PFAS? 2022. Agency for Toxic Substances and Disease Registry (ATSDR). https:// www.atsdr.cdc.gov/pfas/health-effects/overview.html (Accessed July 18, 2023).

Section II

Making Greener Products

4 Greener Product Design Examples

Al Iannuzzi with contributions from Ora Kalaj

There are a lot of companies developing greener products and marketing the greener aspects of these products. Businesses boast of significant sales from these eco-innovative goods. To understand the best techniques and processes used to make greener products, we will evaluate the practices of leading companies; in the end, we will determine which elements are the most common for success. I chose the companies in this study based on their notoriety for exemplary greener product programs; these are the firms I have looked to for benchmarking purposes during my career. Some of the companies that have outstanding greener products that were evaluated using case study analysis are Ecomagination, Patagonia®, SC Johnson GreenList™, Clorox Green Works®, Philips, Samsung Electronics, Apple, Inc., Seventh Generation, Method, Procter & Gamble (P&G), Unilever, BASF, Johnson & Johnson, and Reformation.

ECOMAGINATION

Perhaps the most prominent and successful greener product initiative is GE's Ecomagination. Ecomagination has shaped the way in which the company conducts business, with much of the program's components and motives still in play today. GE is not as vocal about Ecomagination as when it first began but its concepts are embedded in the company's innovation programs. Nevertheless, it will be valuable to evaluate GE's program to learn what the key elements of a successful greener product development program are. We also will get a perspective on how a firm with diversified products, from microwaves and dishwashers to medical imaging equipment to windmills and locomotives, addresses sustainable design. Ecomagination is a well-rounded top-down initiative that has been given significant attention by GE management. When Ecomagination was first released, they used television commercials, print advertisements, and digital marketing to communicate their greener product offerings. They tagged products with improved environmental performance as **Ecomagination**, distributed reports and brochures, maintained a dedicated website, and the CEO at the time publicly spoke about the financial and environmental benefits of this initiative.

DEVELOPMENT OF ECOMAGINATION

When it was first developed, the sustainability-consulting firm GreenOrder was used to help develop the criteria by which to judge products, and a corporate team was established consisting of legal counsel, environmental health and safety, and

DOI: 10.1201/9781003441939-6

marketing representatives to evaluate which products should go into the portfolio. GreenOrder would verify product information and the marketing claims to substantiate the environmental benefits of products. The characteristics considered to designate a product as Ecomagination include energy use, greenhouse gas emissions, water use, and the ability to offer financial benefits to their customers (Iannuzzi 2012). To lend further credibility to GE's greener product approach, an **independent advisory board** was set up. The board consisted of members from non-governmental organizations (NGOs) like the World Resources Institute (WRI), Ceres, and academic institutions, Massachusetts Institute of Technology, and University College of London. Through my research of this program, I have not been able to find anything that describes the methodology used to determine how a product meets the Ecomagination standard. However, as we will see later in the description of some of the products, there are obvious environmental benefits (Iannuzzi 2012).

The greatest strength of Ecomagination is that it is represented as a business initiative, not an environmental one.

In 2017, GE verified product information and the marketing claims to substantiate the environmental benefits of products through external advisors such as the Executive Director of the Institute for Human Rights and Business, the Director of Corporate Social Responsibility of the Harvard Kennedy School, and the Senior Fellow of the Global Green Growth Institute and International Sustainability Development. A Sustainability Steering Committee that consists of leaders across the company with subject matter expertise also oversees the program (GE Annual Report 2015). Perhaps the greatest strength of this program is that it represents a business initiative, not an environmental one. GE described this program in business terms, profits from the products sold and customer desires are met. "Ecomagination is a business initiative to help meet customers' demand for more energy-efficient products and to drive reliable growth for GE." There is a focus on helping customers and society meet the environmental challenges of the day. By investing in cleaner technology and business innovation, GE is committed to driving economic growth and reducing their overall environmental impact. However, GE made it clear that addressing these issues is not an altruistic endeavor; in meeting these challenges, business units must generate "profitable growth for the company." **Positioning this program as a business imperative almost guarantees its success** (GE 2016). I like how GE speaks about this program and how it is weaved into their business processes. "From investing in cleaner technology to developing strategic partnerships to solve some of the toughest environmental challenges, sustainability is embedded in our DNA. **Ecomagination is GE's growth strategy** to enhance resource productivity and reduce environmental impact at a global scale. We are committed to creating a cleaner, faster, smarter tomorrow" (GE 2023).

It's interesting to see that Ecomagination is positioned as meeting the world's need for energy efficiency. According to GE, the world's energy infrastructure must be transformed because it is obsolete and uses environmentally dirty technology.

Cleaner, more reliable, and efficient solutions to support the energy needs of future generations are required. That's where Ecomagination products come in.

ECOMAGINATION PRODUCT CHARACTERISTICS CONSIDERED INCLUDE

* Energy use
* Greenhouse gas emissions
* Water use
* Ability to offer financial benefits to their customers.

Ecomagination was launched in 2005 and increased in size and scope since its beginning. GE has made significant funding investments to demonstrate their commitment; they invested $2.3 billion in clean technology for research and development in 2015. A total of $17 billion has been invested in research and development between 2005 and 2015. They also committed another $8 billion for R&D over the next four years. Success is measured in dollars; in 2015, Ecomagination technologies and solutions generated $36 billion in revenue. Not only are there goals for Ecomagination product sales, but it also sets higher growth targets for these products at double the rate of the overall company growth. This is a huge testimony to the benefits of making greener products.

Setting growth targets that are double the rate of other products in the portfolio breaks the old paradigm that greener products are barely profitable. In fact, Ecomagination products were sold in 100 countries, and GE sees a growing demand across the world. As of 2015, the company has reached 12% reduction of GHG emissions and 17% reduction of freshwater usage. GE continued to invest in this program, setting 2020 goals of $10 billion in additional R&D spend, 20% greenhouse gas reduction, and 20% freshwater reduction. In 2020, GE reported to have spent 3.8 billion in R&D and have reached their goals, achieving 21% greenhouse gas reduction, 21% freshwater reduction. It is evident the company remained committed to their greener product commitments and continues to offer products with enhanced environmental benefits (GE Sustainability Report 2021).

Over a Decade of Ecomagination Results (2005–2015)

$17B R&D Spend	$232B Revenue generated	12% Green-house gas reduction	17% Reduction in water use	40 GW Clean energy installed	1B Gal/Day Wastewater treated	$98 Million Fuel savings

Source: GE Ecomagination Ten Years Later (2015).

The best way to understand the effectiveness of this program is to evaluate a few examples of the products having achieved the designation. The world is constantly in search of more efficient renewable energy sources. GE engineers have been working on the world's most powerful gas turbine, which is also known as 9HA. This machine can easily outdo a steam turbine as it converts natural gas into electricity at more than 64% efficiency. This new technology also transmits analytical information to GE's Predix platform through hundreds of sensors found on the turbine. The 9HA is so efficient that countries around the world are adapting the new technology where in 2021 the Malaysian Sultan Ibrahim Power Plant generates 1,440 MW expected to power 3 million homes (9HA Gas Turbine 2023). Software was also an Ecomagination product; the Predix platform is the world's first industrial app that will monitor and control analytical data from various industrial categories. Data are extremely important when trying to make technology more efficient; the app provides real-time streaming of different machines such as the 9HA and locomotives produced by GE. This platform will provide engineers with an opportunity to create apps for "aviation, agriculture, health care, manufacturing and transportation" and to connect vendors and customers to receive timely analytical results, which results in more efficiency and thus lower environmental impact (GE Get Your Software Kicks on Predix 2016).

Starbucks, for example, was looking for more viable, energy-efficient lighting for their stores. In a partnership with GE, highly efficient LED lighting was developed. As of September 2010, over 7,000 stores have installed the LED lighting; compared to typical store lighting, this saves approximately 8,100 metric tons of CO_2 emissions, the equivalent of 1,600 cars on US highways. Each LED light that is used by Starbucks reduces energy costs and CO_2 emissions by about half a barrel of oil (How Starbucks Saves Millions 2016).

In 2021, GE Healthcare launched SIGNA Hero, a new 3.0T magnetic resonance imaging (MRI) system to provide the global health community with better image quality, faster examination times and sustainable technology. Saving health-care institutions from high operational costs, the new SIGNA Hero decreases the quantity of helium used by 67%, making it 1.4× more efficient than past products (GE Healthcare 2021). Whenever you can position a product with improved operational aspects, e.g., speed, and better environmental performance, it's a winning proposition.

KEY ATTRIBUTES OF ECOMAGINATION

- Represented as a business initiative.
- Strong CEO support.
- Supported with billions of R&D dollars.
- Third-party verified.
- Has review board of external company advisors.

As we can see from reviewing some of the Ecomagination products, there are obvious environmental improvements produced under this banner. The initiative's effectiveness is exemplified in the breadth of the products that have achieved the designation and the deep link to business benefits and integration into the innovation process. Ecomagination is one of the earliest successful greener product design programs, and we can learn a lot from their approach.

BASF

Can a chemical company make greener products? When analyzing one of the largest chemical companies in the world, BASF, we can see how this is possible even in the chemical industry. BASF is a large chemical company based out of Germany, with sales of 87.3 billion euros (BASF About 2023), with 111,000 employees, and has production sites in about 91 countries worldwide. The main products sold are chemicals, performance chemicals, plastics, coatings, catalysts, crop technology, and oil & gas. Tag lines which have been used by the company, "We don't make a lot of the products you buy; we make a lot of the products you buy better®" and "We create chemistry," speaks about the nature of the company. As a chemical company, it's critical to focus on the safety and environmental impacts of your products. BASF believes this so much that they even embedded sustainability into their mission: **"We create chemistry for a sustainable future."** Even one of their market success principles is to **"drive sustainable solutions"** (BASF Confirms 2022). That's one of the biggest drivers for having a greener product development program, **delighting your customers** by helping them with their sustainability goals! I can tell you that at Estée Lauder Companies we look to suppliers of raw materials like BASF to help us achieve our sustainability goals. Having a greener product program that positions your products as having lower environmental impact than the competition is a really smart business decision.

GLOBAL PRODUCT STRATEGY

Being a member of the International Council of Chemical Associations (ICCA), BASF follows their Global Product Strategy (GPS). The GPS is a voluntary standard to ensure chemical manufacturers are following adequate product stewardship practices. The GPS four pillars are as follows:

- *Product stewardship*: The responsibility to understand, manage, and communicate the health and environmental impacts of chemical products at each point in their life cycle.
- *Risk assessment and risk management*: The scientific evaluation of a chemical's hazards, uses and exposures to determine the probability that it will cause adverse effects under real-world conditions, which determines if steps are needed to reduce the risk of harm or misuse.
- *Transparency*: Providing health, safety and environmental information about chemicals to stakeholders and the public to enable customers, regulators and consumers to understand how chemistries can and should be used safely.
- *Stronger chemicals management globally*: Promoting risk-based chemicals management through a mix of government rules, voluntary industry programs, and publicly-sponsored training and recognition programs, particularly in countries without robust chemical management systems.

(BASF Global Product Strategy 2023)

Being transparent about their progress, BASF committed to making their risk assessments available to the public and to generate annual reports that detail their progress. The reports give an overview of the work and collaboration with national and international companies in making sure BASF correctly handles its chemicals in the safest way possible (BASF Risk Management Process 2023).

Eco-Efficiency Analysis

BASF has been an early adopter of **life cycle thinking** using the concept of eco-efficiency which harmonizes the two concepts of economy and ecology. Eco-efficiency is an evaluation of the entire life cycle of a product or process from raw materials sourcing, product manufacture and use, to disposal or recycling. It also considers the consumption methods of the product and its end-of-life recycling or disposal activities. BASF started its work on this process in 1996. They verified the validity of their approach through two independent third-party organizations, TÜV (German technical inspection and certification organization) and NSF (National Sanitation Foundation). The purpose of eco-efficiency is to contrast different materials environmental impacts across their life cycle.

The first step in their process is to evaluate environmental impacts in nine areas:

- Raw materials consumption
- Water consumption
- Land use
- Human toxicity potential
- Eutrophication
- Acidification
- Ozone depletion
- Photochemical ozone creation
- Climate change

Costs incurred in manufacturing and product use are included in the evaluation. Economic and ecological data are plotted on a graph, which is the eco-efficiency portfolio. Figure 4.1 shows the comparison of a product or process with another product or process to show potential ecological and economical improvements.

For ecological evaluations, this analysis is based on DIN EN ISO 14040 and 14044 which has been the standard since 2012. Third parties, the German Association for Technical Inspection and the US National Science Foundation (NSF), are used to validate the Eco-Efficiency analysis, a best practice for greener product development processes. To better understand how this evaluation assists in generating more environmentally friendly products, we can evaluate a few examples (BASF Eco-Efficiency Analysis 2023).

Eco-Package

There is a need for safer and more efficient packaging than the traditional use of plastics from companies and consumers. BASF designed a certified **compostable**

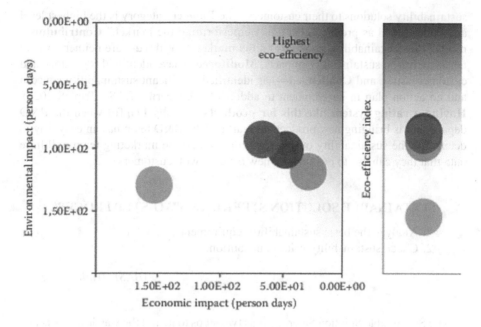

FIGURE 4.1 Eco-efficiency portfolio.

bioplastic. The innovation resulted in a new design called the **Ecovio**. This product is used for plastic films for produce and organic waste bags and agricultural films. BASF is also extending Ecovio's scope to include coating on paper and boards like industrial compostable grade food packaging. BASF received international compostable certification on Ecovio from the European Standard, Australian Standard, Japan Standard, American Standard, and Canada Standard, among others (Ecovio® 2023). Having a certified compostable film for a product that is typically landfilled is an obvious environmental improvement.

ULTRAMID® BIOMASS BALANCE (BMB)

Ultramid Biomass Balance is "virgin-quality polyamide derived from renewable raw material." Polyamide is a group of synthetic polymers that join together to make textiles like nylon. BASF's new innovation will help companies like those in the fashion industry to cut fossil resources by 50% compared to the standard Ultramid. The product is also verified and approved by a third-party organization, REDcert, Ultramid® Biomass Balance (Ultramid BMB 2023). Many companies are trying to reduce their reliance on fossil based raw materials—so this innovation will resonate with BASF's customers.

SUSTAINABLE SOLUTION STEERING®

BASF's greener product development platform is called Sustainable Solution Steering®. This process puts all products into five categories tied to providing sustainability solutions to their customers. The **Pioneer** category is the highest level and is described as products that are "outperforming the market." **Contributor**— meets basic sustainability standards on the market, **Standard**—are neither positive nor negative on sustainability impacts, **Monitored**—have identified regulatory and customer issues, and **Challenged**—has identified significant sustainability concern and an action plan in development to address the concern" (BASF TripleS 2023). **Having a rating system like this for products is really helpful** when the R&D department is bringing new products to market. The R&D team has an easy way to determine the sustainability of the new product, and the marketing team will have data that they can use to position the new product with customers.

SUSTAINABLE SOLUTION STEERING® TWO-STEP PROCESS

1. Analyze the basic sustainability requirements.
2. Check sustainability value contribution.

(BASF TripleS 2023)

The Sustainable Solution Steering® has two steps to it, and the way it works is that the value chain of a product is evaluated from cradle to grave considering industry and region-specific views in markets. Their efforts strive to equally cover their three defined aspects of sustainability: environment, society, and economy. **The idea is to increase the portfolio of innovative and sustainable solutions (accelerators)**, making their customers more successful. Anytime you have a goal like that, it's a pretty good indicator that this is a strong greener product development approach. In 2021, BASF achieved their 2025 targeted goal of 22 billion euros (four years ahead of schedule) of having sales of 24.1 billion euros with Accelerator products, now known as Pioneers. The Triple S method has been followed for more than 10 years, analyzing about 45,000 solutions globally (BASF TripleS 2023).

The criterion for their highest level of classification "pioneers" is that each product must contribute to at least one of the following areas:

- Circular Economy/Resource Efficiency
- Climate change & Energy
- Health and safety
- Biodiversity
- Zero Hunger & Poverty
- Pollution reduction
- Water Protection

(BASF TripleS 2023)

As an example of how Sustainable Solutions Steering works, let's look at how BASF continues to address the concern of poly-fluorinated substances (PFAS). European authorities consider them to pose low risk to human beings and the environment, but BASF anticipated that based on trends, there will most likely be stronger regulation in the future, which turns out to be true. Therefore, PFAS were rated as Challenged under the Sustainable Solution Steering® method when used in paper coatings. They decided to no longer sell these substances and, instead, developed paper coatings that do not accumulate in the environment and are biodegradable (ecovio®) or recyclable (Epotal®) that have been classified as Accelerators (Staying Ahead of PFAS 2023). To give further insight into the type of products that are considered Accelerators, consider the category of products BASF sells to the automotive industry. Products like lightweight plastics which help cars achieve better fuel economy, fuel additives that improve the performance of cars, and catalysts which reduce pollutants are all Accelerator products.

BASF PRODUCT STEWARDSHIP PROGRAM

- Committed to product stewardship through participation in the voluntary GPS.
- Use their own developed Eco-efficiency tool to improve the sustainability of their products.
- Have a greener product development approach called Sustainable Solution Steering® which classifies products into five categories: Pioneer, Contributor, Standard, Monitored and Challenged.
- Use of third-party evaluations and certification to endorse the improvements of products.

BASF being a large chemical company has made assurances to their customers and stakeholders that their products will be both safe for human health and the environment. Commitments have been made to conform to voluntary international standards to ensure products have adequate performance. To foster lower environmental impacts of their products, BASF has developed an eco-efficiency tool that measures the footprint of their products and cost benefits. This tool has been used to demonstrate the improved benefits of new products, helping to meet customers' greener products desire. A greener product approach called Sustainable Solutions Steering is used to drive products into higher levels of performance, and goals have been set to increase their portfolio toward the highest sustainability rating. The BASF case study clearly demonstrates that chemical companies can develop greener products and it has clear business benefits.

PATAGONIA

In 1973, Yvon Chouinard's deep admiration for nature led him to make clothing and gear for his friends creating Patagonia. In 2022, the company's revenue hit approximately \$1.5 billion, where CEO, Yvon Chouinard, donated 98% of the company to Holdfast Collective, a trade-based organization. The company in charge will donate its profits to combat the environmental crisis and defend nature, not your typical for-profit company approach (Chouinard 2023). Patagonia developed an impressive greener product development program in the outdoor recreation clothing sector. The company created a wide range of environmental responsibility programs to tackle the problems created by the apparel industry. According to Quantis, a leading sustainability consultancy, the **fashion industry contributes to "6.7% of the global greenhouse gases emitted in the world** and releases 2–3.29 billion tons of carbon dioxide emissions." In doing its part, Patagonia has achieved 100% renewable energy for their owned/operated stores, offices, and distribution stores, and they are currently at 95% renewable energy from material manufacturing operations (Patagonia Our Environmental Responsibility Programs 2023). When speaking of raw materials, growing agricultural raw materials utilizes harmful techniques and chemicals that impact the environment like pesticides and herbicides. In 1996, Patagonia decided to use organically grown cotton and increase their usage of preferred materials: organic and regenerative organic cotton, hemp, recycled polyester, recycled nylon, and more (Patagonia Organic Cotton 2023).

Sustainability is a journey, and Patagonia recognizes that there is always more they can do. The CEO of Patagonia, Ryan Geller stated, "At Patagonia we have to be greener than green – and make a profit" (Balch 2023). Patagonia has set the three following climate goals that they will be working toward:

- By 2025, eliminate virgin petroleum fiber in products and only use preferred materials.
- By 2025, packaging to consist of 100% reusable, home compostable, renewable or easily recyclable.
- By 2040, achieve net zero* across the entire business.

(Patagonia Our Environmental Responsibility Programs 2023)

Patagonia prides itself in research and innovation, looking for best possible methods for all materials that go into Patagonia's line. Scientists and engineers use data to develop, test, and improve materials to help reduce the company's environmental impact. But make no mistake such work is no easy feat, the research takes time as you must repurpose the materials in an optimal approach. The R&D Team looks to steer clear from regrettable substitution: a harmful chemical that is banned and is replaced by a new chemical that is just as hazardous. The following materials exemplify Patagonia's ongoing commitment to the environment without compromising the quality of its clothing and equipment (Preda 2023).

At Patagonia we have to be greener than green—and make a profit.

Ryan Geller
Patagonia CEO

Regenerative Organic Certified Cotton—Industrial agriculture is responsible for 25% of global emissions. Aware of the detrimental effects, Patagonia has partnered with the Regenerative Organic Alliance to help produce eco-conscious materials like cotton. The certification ensures the highest standards of organic agriculture in the world, having requirements on soil health, animal welfare, and social fairness. The practices that make up regenerative organic agriculture focus on ways to sustainably produce for the future generations. These practices include cover cropping, crop rotation, low-to-no-till, compost, and zero use of persistent chemical pesticides and fertilizers. In 2018, Patagonia piloted the program working with 150 farmers to achieve the standards of the certification. In 2023 the program worked with over 2,200 farmers to launch the first Regenerative Organic Certified Cotton. The company does not plan to stop with cotton but will look to work toward certification with other materials as well.

Advanced Denim—The denim manufacturing industry is responsible for chemical pollution/contamination, high energy/water usage, and unfair labor treatment. The dyes and chemicals that go into making the infamous denim blue are responsible for much pollution in the world. "The denim industry uses more than 45,000 tons of synthetic indigo a year with over 84,000 tons of sodium hydrosulfite as a reducing agent and 53,000 tons of lye," to allow for the adherence of indigo on fabrics (Beeson 2021). To move away from indigo, Patagonia has launched Advanced Denim Technology using sulfur dyes; the new technology "uses 84% less water, 30% less total energy, 50% less electricity and produces 25% less CO_2 emissions." Although the company has made impressive accomplishments, they continue to research new and eco-friendly dye technologies.

Yulex Natural Rubber—In 2016, Patagonia replaced the neoprene rubber, traditionally found in wetsuits, with natural rubber to reduce the use of petroleum-based materials. Created in 1930, neoprene, a synthetic non-renewable rubber, was made by the chemical processing of petroleum, a by-product of crude oil. In attempts to alleviate the unsustainable processes, companies looked to limestone as the new alternative. However, the extraction and production of such material requires massive amounts of energy, continuing the line of harmful practices. Patagonia became the first ever surf company to go **neoprene free** by utilizing the **natural rubber** found in rubber trees (hevea trees). The company partnered with Yulex, a first ever producer of natural rubber latex in America, to design more sustainable and better wetsuits. As hevea is a cash crop, Patagonia partnered with the Forest Stewardship Council to ensure its business contributes to upkeep the integrity of the ecosystem. It is important to note that although the wetsuit is a greener product, **it simultaneously does not compromise performance**, with better warmth and flexibility than neoprene. Patagonia is also planning to use this innovative greener raw material for its gloves, boots, and hoods.

KEY ATTRIBUTES OF PATAGONIA'S PROGRAM

- Shift away from petroleum based raw materials.
- Commitment to regenerative agriculture for raw materials.
- Focus on the circular economy with their Worn and Wear program.
- Using their business to solve environmental problems.
- Use of innovative lower environmental impact materials.

Biobased Polyester—Similar to the chemistry of neoprene, polyester also originates from petroleum where its components are distilled into different chemicals and then heated to create the plastic-based substance. The UN estimates that 73% of clothing globally will be made from polyester, nylon or acrylic continuing the reliance of Big Oil companies. In 2014, Patagonia began to look for ways to substitute polyethylene terephthalate (PET) that will decrease their environmental impact.

NetPlus Recycled Fishing Nets—The motive to repurposing fishing nets for their clothing line is the following staggering statistic: "8.8 million tons of mismanaged plastic enter oceans every year, most of it single-use." To help solve the issue of abandoned fishing nets in the water, Patagonia partnered with Bureo, a California-based company, working with local fishers in South America to recycle the nets into **NetPlus, a 100% fully traceable post-consumer material**. NetPlus is used in a wide range of products from clothing to equipment and accessories. Through the program, Patagonia **collected more than 2,000 MT of fishing nets** from the ocean and for the fall 2023 season will incorporate 185 MT in their clothing line.

Outside of material innovation, Patagonia has also addressed sustainability through their **circularity program, Worn and Wear**. In 2018, the US Environmental Protection Agency (EPA) stated that **85% of clothing ends in landfills or gets incinerated**. The initiative engages customers to trade in their old Patagonia gear and buy the brand's used clothing (Worn & Wear 2023). Worn and Wear keeps Patagonia products in circulation for the long run and generates $5 million annually (Balch 2023).

Patagonia is a pioneer in developing greener products and a leader in sustainable innovation of new lower impact raw materials. Having sustainability built into the foundation of the company enables great performing greener products that their customers love. When your founder says "**Earth is our only shareholder**" it's a strong sign that the products developed will embody the highest level of environmental protection (Balch 2023).

SC JOHNSON GREENLIST™

SC Johnson is a privately held company based in Racine, Wisconsin, and is a global provider of consumer products. With over $11.8 billion in sales, their product lines include a variety of merchandise used in homes, such as Glade air fresheners, Ziploc plastic sandwich bags, household cleaners Pledge and Windex, and even the Raid pesticide brand. Manufacturers of these product groups have been challenged by customers and NGOs to improve their formulas because of questionable ingredients that may have toxic effects on the environment and to the customers that use them. The market has been shifting with the introduction of newer products claiming to be more natural or greener.

SC Johnson has been very public about focusing on greening their products. To positively affect product design, they **made sustainability a company-wide initiative** and not a single department's responsibility. The goal of creating the green list was to focus on making "better" options for products instead of focusing on taking out the "bad" ingredients. The primary way that products are greened is by continuously improving the raw materials used. The measurement system to drive the greener product improvements is a process called Greenlist™ (Public Sustainability Report 2015). Similar to Ecomaginaton, SC Johnson was an early adopter of a greener product development system called the **Greenlist™** in 2001. This continuous improvement process rates each ingredient from 3 to 0. Ingredients are put into categories of **Best (3), Better (2), Acceptable (1), and the least desirable to use, 0-rated materials**. Each ingredient is based on four to seven key criteria such as **biodegradability and toxicity**. The objective is to increase the score of a product's ingredients over time. The 0-rated materials are a specific area of focus and are only to be used if there is no other workable alternative. Over time, the company created the "Restricted Use" list compiled of 0-rated ingredients, including over 200 unique raw materials and over 2,400 fragrance materials. It is very useful to have tools and objectives to enable product developers to develop greener products like Greenlist™; it makes it easier for positive decisions to be made. SC Johnson scientists are tasked with developing new products that use raw materials rated as Better or Best. As with most consumer product companies, there are times when product formulations are updated; in such a case, the scientists must include ingredients that have ratings equal to or higher than the original formula.

The Greenlist™ approach not only focuses on improving products but also focuses on supply-chain practices. The company puts a higher value on suppliers that demonstrate environmental responsibility through such programs as ISO14001 certification. Also, they have a focus on preventing deforestation through the sustainable sourcing of pulp, paper, packaging, and palm oil. Suppliers that help with this commitment will no doubt gain opportunity for more business (SC Johnson Greenlist Program 2017).

RESULTS OF GREENLIST™

The benefit of having **a metric-based system to rate your products is that progress can clearly be demonstrated to stakeholders** (see Figure 4.2). Using 2000/2001 as a baseline in the reporting year 2015/2016, SC Johnson's product formulas improved the amount of what they consider the best ingredients (3- and 4-rated ingredients). Back in 2001, the company started at 18% "Better/Best" ingredients and in 2016 at 52%, indicating to all concerned that significant progress has been made. The SC Johnson Greenlist™ process score uses a four-step process in examining its ingredients: chronic human health, long-term environmental hazards, acute risks to human and environmental health, and other potential effects.

GREENLIST™ FOUR STEPS TO BETTER

1. Chronic human health
2. Long-term environmental hazard
3. Acute risks to human and environmental health
4. Other potential effects

One of the ways to **generate more public trust** in the environmental and health claims a company makes is to **become more transparent** about a product's ingredients. There have been a lot of NGO reports on the hazardous nature of household products, such as the type manufactured by SC Johnson. Taking a bold step, the

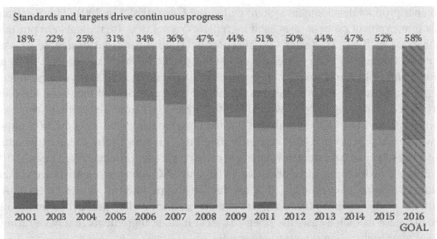

This chart shows SC Johnson's progress increasing the use of Better/Best ingredients as rated by the SC Johnson Greenlist™ process. Due to ongoing advances in measurement, changes in our product portfolio, and increasing numbers of ingredients and materials being measured, data may not always reflect apples-to-apples comparisons year on year. Also, scores after 2011 reflect combined ingredient and package scores. Prior to 2011, scores were ingredients only. Totals are rounded.

The SC Johnson Greenlist™ process uses a four-point scale:
■ 3 - Best   ■ 2 - Better   ■ 1 - Good   ■ 0 - 0-rated materials

FIGURE 4.2   Greenlist™ progress for increasing the use of better and best ingredients.

company has committed to make all of its ingredients available to the public (SC Johnson Greenlist Program 2017). SC Johnson, going the extra mile, has committed to list all ingredients, including fragrances and preservatives, on a dedicated website called whatsinsidescjohnson.com. This is a leadership position for the industry and a trend that I believe will eventually be a requirement. As of 2015, SC Johnson began disclosing the fragrances used in products and is committed to complete and total transparency of their products' ingredients (SC Johnson Greenlist Program 2017). In 2017, SC Johnson Greenlist™ made available their ingredient information to more than 5 billion consumers in 52 countries worldwide. To help increase transparency and accessibility, the company strategically provided their ingredient information in a total of 34 languages on more than 5,300 products (SCJ Sustainability Report 2017).

KEY ATTRIBUTES OF SC JOHNSON'S PROGRAM

- Greenlist™ metric-based ingredient improvement system.
- Committed to make all product ingredients available to the public, including product-specific fragrances.
- Connect product sales to worthy causes to promote human health.
- Focus on making products available to the bottom of the pyramid.
- Speak about dilemmas publicly.

BASE OF THE PYRAMID

SC Johnson makes a concerted effort to source and sell products to the poorest individuals in the world, which they call the base of the pyramid. They are trying to bring forth products that are affordable and source products that can bring a social benefit to the poorest people in the world. This is admirable and a great example of how business can be a force for good—by making products available to the neediest.

An example of this is sourcing the pest control product pyrethrum from local suppliers in east Africa. Pyrethrum can be manufactured through a synthetic process; however, it can be derived naturally by extraction from chrysanthemums. To enable economic growth and a sustainably sourced raw material, SC Johnson is working with local suppliers in East Africa through a partnership with the US Agency for International Development to improve farming techniques which could enable a more consistent supply. Through this effort, farmers in Africa are able to increase their crop yields and make more profit while providing a naturally derived ingredient. New supply chains are being developed to bring products to those that traditionally could not afford them. The SC Johnson Company created the Spatial Repellent Program to create effective pest control products for underserved communities to help reduce insect-borne diseases like the Zika virus. In 2014, the company developed the Mosquito Shield that is said to be light-weight, cost efficient, and easy to use in all regions across the world. Up to the year 2021, Indonesia experienced 66% reduction in malaria infection overall, Peru experienced 34% in Aedas-borne virus infection rates, and Tanzania experienced 69% reduction in bites from highly pyrethroid-resistant Anopheles Arabiensis.

The SC Johnson Hygiene Program helps to serve under-resourced populations in need of hygiene care to help combat targeted diseases. In 2020, SC Johnson Base of the Pyramid Group, United Nations Foundation, and the Mentor Initiative sent disinfectant kits to 100 health facilities serving 2 million residents in Nigeria's Borno State. Between 2013 and 2021, the SC Johnson company reached over 71.5 million people and funded over $30 million through their program. At SC Johnson they often say, "**We don't measure our success by the financial results of the next quarter, but by the world we leave for the next generation**." The Base of the Pyramid group will continue to support future generations to come by helping bring products and initiatives that will minimize disease (Base of the Pyramid Impact Report 2021). Helping the most vulnerable people of the world with your products is a really great way to demonstrate your company's values.

SPEAKING ABOUT DILEMMAS

NGOs have focused on low-level concentrations of potentially problematic ingredients in consumer products. Publicly discussing a point of view and sharing data can build trust in your product stewardship program. One of the ingredients of concern which is used in some of SC Johnson's products is phthalates. The one that has received the most negative focus is diethyl phthalate (DEP). This issue is discussed on their website, where they state that the evidence of numerous studies has indicated that DEP is safe for use, as is the whole class of phthalates. But since the company is committed to going above and beyond, they have decided to phase out the use of DEP in all their fragrances.

Anyone who has worked in the area of product sustainability will know that **perception is reality** in the eyes of the public. Conceding this fact, SC Johnson explains that although the scientific evidence indicates that there is no concern, they understand the consumer's desire to have products without materials of concern. So, suppliers were asked to phase out DEP in all their fragrances. Speaking out about dilemmas such as this builds public trust and lends credibility to discussions with NGOs. Sometimes difficult decisions have to be made based on customers' perceptions, even if company scientists disagree. **Only companies that are leaders in product stewardship are willing to discuss sensitive issues such as product ingredients, publicly**.

SC Johnson deploys a metric-based system that makes it easy for developers to focus product improvements and to report public progress. They have also connected their products to special causes that will benefit the poorest consumers. There is a commitment to greening up all their product lines through their Greenlist™ process. SC Johnson has taken a leadership position in voluntarily listing all product ingredients on their website. Many of the concepts used in the development of the Greenlist™ and SC Johnson's greener product approach give us significant insight to what it takes to have an effective product sustainability program.

CLOROX GREEN WORKS®

Clorox, best known as the makers of bleach, also markets such products as plastic wrap, household cleaners, water filters, and charcoal for grilling food. In recent years, they have become well known for developing eco-innovative products. The most meaningful way to improve a product's environmental performance is to focus on its greatest life-cycle impacts. When looking at consumer goods, manufacturers find that the most significant impacts are in the product itself not in the manufacturing plant. This is especially true when you place over 1 billion products in the marketplace every year.

An excellent example of developing and marketing a greener product came as a surprise to many. The entry of the Green Works® line was revolutionary and a real game changer for the household cleaning industry. Clorox, which is known for creating bleach and various other successful biocides, re-branded their image as safe and free of harsh chemicals. Green Works® products are based on a natural ingredients' platform. Perhaps the greatest accomplishment of Green Works® is not that they have had so much success, but more in the way **it vaulted greener products into the mainstream marketplace**. The old stereotype of green products being ineffective and not of high quality was broken. Having a leading manufacturer of cleaning products introduce a natural product line enabled the public to embrace and feel safe purchasing greener products. Clorox explains that the reason Green Works® products were developed was to meet customers' demand for naturally derived cleaning products. Their evaluation of the market was right on and apparently their timing was perfect. In my opinion, the success of the Green Works® line paved the way for other brands, even those that were not previously sold in the mainstream grocery store, such as Seventh Generation and Method.

GREEN WORKS PRODUCT STANDARDS

- Made with plant and mineral-based cleaning ingredients.
- Come from biodegradable ingredients that are naturally derived.
- Are not tested on animals.
- Use sustainable packaging whenever possible.
- Are acknowledged by the EPA's Safer Choice program.

CERTIFICATION AND PARTNERSHIP

When Green Works was first developed, Clorox took two significant steps to bolster the confidence and trust of customers and stakeholders—receiving the US EPA Design for the Environment (now called Safer Choice) product certification and getting the endorsement from a major environmental NGO, the Sierra Club. By doing this, their new green brand's "natural" image was reinforced (Iannuzzi 2012). A smart move for Green Works® was partnering with the Sierra Club when it was introduced in 2008. This backing is significant because the Sierra Club is one of the best-known environmental groups in the world. Products displayed the Sierra Club logo which indicates its endorsement of the brand. The Sierra Club endorsement is not used by Green Works anymore, but using their status to endorse their new line of green products at the time was a very smart, strategic marketing technique that gave the brand credibility.

The Safer Choice certification requires a stringent evaluation of the product's ingredients. Products that receive this certification have reached a level that indicates their ingredients have the lowest impacts in their product class. It is designed to protect the health of families and the environment. Though not widely recognized by consumers, the use of the Safer Choice logo on a product does go a long way toward defusing criticism by NGO watchdog groups and makes retailers more comfortable with touting the product as greener. Since the EPA uses a robust process of ingredient reviews and the use of predictive models, their certification quells claims of greenwashing.

An additional natural line of products Clorox acquired in 2007 is the iconic Burt's Bees®. Sold in mainstream supermarkets, it is a leading natural personal care brand in the United States that is also marketed internationally. Clorox has expanded Burt's business model, one that has a strong social responsibility component, where they give 10% of all sales to worthy causes. In Burt's Bees® 2020 Impact report, 100% of their packaging is recyclable, utilizing an average of 50%

recycled materials in packaging, and is CarbonNeutral Certified (Burt's Bees Impact Report 2020).

CLOROX INTEGRATED IGNITE SUSTAINABILITY GOALS

Transparency Commitments
- Increase sales of products with targeted certifications and product transparency affiliations 100% by 2025, compared to 2020 baseline.

Ingredient Management
- Improve Chemical Footprint Project survey score for our domestic cleaning portfolio 50% by 2030 with an interim target of 35% by 2025.
- Publicly share C restricted substances list for domestic retail cleaning products by calendar year 2020 and additional categories by 2025.

Plastics & Other Wastes
- 100% recyclable, reusable or compostable packaging by 2025.
- 100% zero waste to landfill in global facilities by 2030; plants by 2025.

Climate
- Achieve science-based targets to reduce greenhouse gas emissions by 2030 versus a 2020 baseline.
- Achieve net-zero greenhouse gas emissions across scopes 1, 2 and 3 by 2050.

(Integrated IGNITE ESG Goals 2023)

Green Works® and Burt's Bees® have been a big success, but **what about Clorox's legacy products?** To address their main products, Clorox adopted the IGNITE Strategy in 2019, to promote innovation in different areas of business and integrate environmental, social, and governance decision making. The information above are corporate level goals tackling sustainability initiatives under the IGNITE Strategy Program. Clorox's legacy products have completed their own life-cycle assessment (LCA) and set specified ESG objectives (Integrated IGNITE ESG Goals 2023). This is a common company greener product approach, having enterprise-wide approaches to all products and select lines of deeper green products.

Examples of greening existing product lines include the technological development of the Glad® kitchen drawstring bags (trash bags) having a competitive edge over the top ten companies in the market with 7–22% less plastic. Shifting to Glad® ForceFlexPlus Recovered Materials Trash Bags, the company uses 20% of diverted post-consumer recycled plastic, 30% of plastic from reused material in manufacturing processes, and 100% recycling packaging. In the Australia/New Zealand sector, the 50% Plant-Based Cling Wrap, Sandwich and Snack Reseal Bags are made from a renewable resource, sugar cane (Clorox Products & Brand Packaging Highlights 2023). Having products that are inherently greener is another way to demonstrate a company's commitment to sustainability, enabling customers to reduce their environmental impacts. The **Brita® water pitcher** filter provides clean drinking water. Customers using this instead of disposable bottles **can replace as many as 900 standard 16.9 oz. bottles of water per year**. This is good not only for reducing plastic bottles but also for saving consumer's money. Clorox estimates that a Brita® pitcher filters 240 gallons of water per year for about 19 cents per day. This is a substantially lower cost than purchasing a bottle of water. The brand also allows consumers to return their used Brita filters and pitchers into 100% recycled products through their take-back partnership with TerraCycle (Clorox Products & Brand Packaging Highlights 2023).

KEY ATTRIBUTES OF CLOROX GREEN PRODUCTS INITIATIVES

- Development of the natural products line Green Works®.
- Use of concentrated formulas to reduce impacts.
- Transparency of ingredients used in products.
- EPA's Safer Choice to bolster green credentials.
- Set sustainable product and packaging goals for legacy products.

A big impact area for consumer goods is packaging waste. Many Clorox company brands have moved to more sustainable packaging. Package redesign, material reduction, and increases in recycled content have led to measurable improvements. Clorox has made significant progress in this area; the Clorox brands have introduced concentrated formulas for Clorox Bleach and Bleach Powder to help decrease the products' carbon footprint. The **concentrated formula** for Clorox Bleach saves 23 million gallons of water, 7.4 million pounds of paper equivalent to 64,750 trees, and 15 million pounds of plastic annually. The concentrated formula for Clorox Bleach Powder uses 27% post-consumer recycled content, 50% less plastic, and 90% less water. In 2019, Clorox launched the revolutionary **Compostable Cleaning Wipes,** and the product earned the Environmental Protection Agency's Safer Choice certification (Clorox Products & Brand Packaging Highlights 2023).

Another leadership position taken by Clorox was being the first in the cleaning industry to commit to voluntary disclosure of product ingredients. Work on this objective started in 2008 when all ingredients were listed on product labels for the Green Works® line. Traditional brand household and commercial cleaners had their ingredients listed on the corporate website in 2009. Clorox states that their priority is keeping families, pets, and the environment safe.

Clorox has been a trailblazer in bringing greener products to the mainstream consumer with the Green Works® line, a major event in the greener product movement. Not only have more sustainable product lines been brought to market but also legacy product lines are being greened up by setting corporate goals to improve the environmental profile of products and packages. The credentials of the products are bolstered with third-party certifications and donations of a portion of sales to good causes.

JOHNSON & JOHNSON

Johnson & Johnson (J&J) is the world's largest health-care product provider; it is also the company I know a lot about since I worked there for over 20 years. J&J has a very broad line of products, with sales more than $94.943 billion (2023). There are three main divisions: consumer products, medical devices, and pharmaceuticals. Some of the brands that may be familiar to you include Johnsons' baby products, Aveeno, Neutrogena, the Acuvue contact lens, over-the-counter pain medication Tylenol and Motrin, and in addition numerous medical devices like replacement hips and knees, sutures and various pharmaceutical products that treat migraines and rheumatoid arthritis to cancer (note that in 2023 J&J spun off the consumer products division into a new separate company called Kenvue). Developing product steward-ship programs for a wide array of different products has its challenges. I have been fortunate to work for a company that embraced the concept of product stewardship before it became a business imperative. In the 1990s, J&J's first product steward-ship initiatives began with the public facing pollution prevention goals. These goals were primarily traditional footprint reduction goals at manufacturing facilities (e.g., water, waste, energy reductions), but there also was a packaging reduction goal, the first attempt at reducing the impacts of the product itself. In the late 1990s, I was privileged to develop and lead the J&J design for the environment (DfE) program. This initiative covered all aspects of the product life cycle, from evaluation of raw materials to improvements of the manufacturing process, the product and package, emerging issues as well as end-of-life disposal issues. In 2000, the company started to develop and report on public facing sustainability goals every five years. The goals, called the Next-Generation Goals, included a requirement that all products and packages have an environmental impact analysis performed, considering ways in which impacts can be reduced.

EARTHWARDS®

In 2009, the feeling was that the company needed to get more out of the DfE program and make it easier for development personnel to determine how to make a product greener and for marketing personnel to communicate the greener attri-butes of product improvements. This was the rationale behind the development of Earthwards®. Earthwards® is a process that enables product-development teams to evaluate a product throughout its life cycle and identify areas where it can be improved, to lower its impact and increase social benefit. Tools and a scorecard were developed to assist design teams to uncover improvements to reduce prod-uct impacts. Products that have been significantly improved can receive special designation if they complete a life-cycle screen, meet pre-requisites, and achieve the Earthwards® criteria. The scorecard was developed through benchmarking leading companies through interviewing internal and external stakeholders, and by the guidance of a leading product stewardship consultant, Five Winds International. In addition, over the years the program was formally evaluated by experts from government, academia, business, and NGOs to make recommendations for improvements. Earthwards® not only was used to improve product design it was incorporated into the business by **setting an external goal:** "new and existing products representing 20% of Johnson & Johnson revenue achieve Earthwards® recognition for sustainable innovation improvements" (Product Stewardship/Earthwards 2023).

PROCESS STEPS

The Earthwards® approach consisted of four steps.

1. *Meet product stewardship requirements*: New products must achieve regulatory compliance and deliver on Johnson & Johnson's high standards. Product teams are to answer a series of questions; examples include:
   a.   Are materials sourced from environmentally or culturally sensitive regions?
   b.   Identify 100% Watch List materials/ingredients in our products.
   c.   Have you done a review to assess where the product and packaging end up after use?
2. *Be reviewed for life-cycle impacts*: The life cycle impacts of products are reviewed at the category level, and opportunities to drive improvements are considered at the design, procurement, manufacturing, and marketing stages of a product's development.
3. *Implement and validate improvements*: Product teams collaborate with sustainability experts to implement recommended improvements, and environmental marketing claims are reviewed and approved in accordance with applicable guidelines.
4. *Achieve Earthwards® recognition*: an honor celebrating our most innovative and improved products. If a product achieves at least three significant improvements across seven impact areas, a board of internal and external experts determines if the product warrants Earthwards® recognition (Figure 4.3).

(Product Stewardship/Earthwards 2023)

SCORECARD APPROVAL

If the product team believes that at least three sustainable improvements were made within seven categories, then they can submit the scorecard for verification to a review board consisting of internal J&J experts, legal counsel, public affairs, third parties including academics and an NGO. The board then makes a decision whether the product meets the Earthwards® criteria. If it does meet the criteria, the product receives the Earthwards® designation and is considered

FIGURE 4.3 The Earthwards® approach. (Earthwards® is a registered trademark of Johnson & Johnson. Reproduced with permission from Johnson & Johnson.)

a recognized product. We are careful in the words we use to describe our most improved products and through much debate and thought came up with calling them Earthwards® "**recognized**" products. Since this is an internal company program, we felt the word "recognized" would not allow the Earthwards® logo or mark to be confused with a third-party eco-logo. If we used a word like "certified," that could confuse some customers into thinking it might be an eco-logo. As discussed in detail in one of the latter chapters in this book, words matter when trying to avoid being accused of greenwashing.

The seven areas for improvement are:

1. Materials
    a.   Meet consumer needs with less material.
    b.   Use more environmentally preferable material.
2. Packaging
    a.   Reduce packaging.
    b.   Use more sustainable packaging materials.
3. Energy
    a.   Create a less energy-intensive product.
    b.   Use more efficient manufacturing and distribution processes.
4. Waste
    a.   Reduce waste during manufacturing.
    b.   Recover more products for reuse or recycling.

Earthwards® Scorecard		
Pre-requisites		Achieved
	Know materials and ingredients in your product	
	Identify and plan to address J&J Watch List materials	
	Know where product and packaging end up after use	
	Know whether agricultural or mined ingredients come from culturally or environmentally sensitive regions	
Complete life cycle screen to *identify priority goals* for the product (see separate screening questions)		
Goals		
Materials	1 Meet consumer need with less material *or*	
	2 Use more environmentally preferred material (see list) *or*	
Packaging	3 Meet consumer need with less packaging *or*	
	4 Use more environmentally preferred packaging material (see list) *or*	
Energy	5 Make product more energy efficient in use *or*	
	6 Make manufacturing or distribution more energy efficient *or*	
Water	7 Make product more water efficient in use *or*	
	8 Make manufacturing more water efficient *or*	
Waste	9 Make product with less waste during manufacturing *or*	
	10 Recover more product, after use, for reuse or recycling	
Results	**Achieved all pre-requisites + three other goals?**	

FIGURE 4.4 Johnson & Johnson's Earthwards® scorecard.

5. Water
   a. Generate a more water-efficient product.
   b. Make manufacturing process more water-efficient.
6. Innovation
   a. Initiate quantifiable environmental improvements in a product or process that has not been captured in another scorecard category.
7. Social
   a. Use fair-trade materials, select socially responsible suppliers, or support causes with clear social/environmental benefits (Figure 4.4).

(*Johnson & Johnson 2010*)

EARTHWARDS® RECOGNIZED PRODUCT EXAMPLES

Simponi® (Golimumab)

If you were suffering from rheumatoid arthritis, would you prefer your treatment injections to be once per month or three or four times a month? Simponi® is an innovative therapy that requires self-injections you can do in the comfort of your home. Not only is this therapy more convenient and less invasive for the patient but also it is much more sustainable than other treatments, using less materials and less energy. To help further drive sustainability at Johnson and Johnson Simponi implemented the program, Safe Returns, allowing its patients to safely return their devices through an eco-friendly envelope kit (Safe Returns 2023).

Innovation

By requiring only 12 injections per year, patient needs are met using 36–61% less material. In the United States, a new sample distribution system includes a shipper that is returnable and reusable and employs USDA-certified bio-based cooling materials. The previous shipping container was single-use disposable.

Earthwards® has two objectives:

1. Give clear line of sight to R&D and marketing professionals on how to make a product greener.
2. Develop green marketing claims backed by science, facts, and data.

PACKAGING

The new shipper is 50% lighter and helps prevent disposal of more than 42,000 cubic feet of Styrofoam annually compared to the original shipping method (Iannuzzi 2012). This is a perfect example of an Earthwards® recognized product—one that is **better for the customer (the patient in this case) and has improved environmental performance** to a comparison product. When describing Earthwards® I would say that it has two objectives: (1) give a clear line of sight to R&D and marketing professionals on how to make a product greener, and (2) develop green marketing claims backed by science, facts, and data. Some other examples of Earthwards® recognized products include the iconic **Johnson & Johnson Baby Shampoo**. First launched in 1953, the company realized the need for a "low-environmental-impact formula." The product's manufacturing process yielded 71% less water and 35% less energy.

The **Surgicel®** Absorbable Hemostat, a knitted fabric used to help with bleeding during surgery, earned Earthwards® recognition in its packaging and transportation. The box is made of 100% responsibly sourced paper and reduced 57% of the amount of tertiary packaging. With the change in packaging, the product increased shipping space by 62%, meaning less trucks and less greenhouse gases emitted.

ZYTIGA® is used to treat prostate cancer, which is the second most frequently diagnosed cancer in men and the fifth most common cancer overall. By applying green chemistry principles in the formulation of this product, Janssen (a J&J pharmaceutical company) doubled the process yield and realized significant reductions in raw material use, water use, and hazardous waste generation.

The product uses 64% fewer raw materials, 87% less hazardous waste, and 78% less water (Mazur 2017).

HEALTH FOR HUMANITY 2025 GOALS

J&J established Health for Humanity 2025 Goals to focus on three areas that will change the health of humanity: Champion Global Health Equity, Empower Our Employees, and Advance Environmental Health. The following goals will address energy and climate:

- By 2025, source 100% of electricity needs from renewable sources.
- By 2030, achieve carbon neutrality for our operations, going beyond our Science Based Target to reduce absolute Scope 1 and 2 emissions 60% from 2016 levels.
- By 2030, reduce absolute upstream value chain (Scope 3) emissions 20% from 2016 levels.

(J&J Health for Humanity Goals 2022)

EARTHWARDS® RESULTS

Since its inception, Earthwards® has steadily grown, and at the end of 2016, there were 96 of the most improved "recognized products." To demonstrate the impact of the initiative, sales of the recognized products are tracked. At the end of 2015, there was approximately $9.3 billion in revenue—roughly 13% of sales—from Earthwards®-recognized products. Impacts are not only calculated by sales but on environmental improvements: at the end of 2015, approximately 3,600 metric tons (MTs) of reduction in **packaging**, 18,270 MTs of **materials** removed, 6,630 liters of less water used, and 3,630 MTs of waste reduced (Iannuzzi 2017). Improvements in products were made across the seven Earthwards® categories, covering all the categories, as you can see from the list below. Further connecting greener products to business results the company set a 20% of revenue to be generated from Earthwards®-recognized products (J&J Earthwards 2024).

Impact Area	Number of Improvements
Materials	85
Packaging	70
Energy	45
Waste	19
Water	18
Social	35
Innovation	25

Source: Johnson & Johnson, Citizenship & Sustainability
Report, p. 9, 2015

HEALTH FOR HUMANITY 2025 GOALS

J&J has put out public commitments for sustainability goals since the 1990s. New goals are developed every five years, and the current goals are called Health for Humanity 2025 Goals. The goals are set into three categories, Champion Global Health Equity, Empower our Employees, and Advance Environmental Health. The Advance Environmental Health goals cover various initiatives that protect the environmental health and the resilience of the business.

ADVANCE ENVIRONMENTAL HEALTH GOALS

- Renewable Electricity
- Carbon Neutrality for Operations
- Scope 3 Emissions Reductions

(J&J Health for Humanity Goals 2022)

TARGETS AND METRICS

- By 2025, source 100% of electricity needs from renewable sources.
- By 2030, achieve carbon neutrality for operations, going beyond our Science Based Target to reduce absolute Scope 1 and 2 emissions 60% from 2016 levels.
- By 2030, reduce absolute upstream value chain (Scope 3) emissions 20% from 2016 levels.

J&J has also implemented another category for their Health for Humanity 2025 Goals called Lead with Accountability & Innovation. The specified class includes the three categories (Champion Global Health Equity, Empower our Employees, and Advance Environmental Health) to actively change the course of humanity for the better. The company's objectives are representative of their "commitment to science and data-driven decision making, compliance, integrity and responsible business practices across the value chain." The 2025 Lead with Accountability & Innovation targets are as follows:

- By 2025, expand the Johnson & Johnson Supplier Sustainability Program to include all suppliers; monitoring, engaging, collaborating on joint environmental, social and ethical obligations.
- By 2025, achieve $4.5B Global Impact Spend with small and diverse suppliers, representing a 20% increase from 2020.
- By 2025, establish 10 supplier Partnerships for Good, collaborating with our suppliers to create innovative solutions that drive environmental, social and economic improvements

(J&J Health for Humanity Goals 2022)

JOHNSON & JOHNSON GREENER PRODUCT PROGRAMS

- Earthwards® process is used to develop greener products using life cycle thinking and seven key focus areas.
- Earthwards® process generates fact-based customer facing claims.
- Products making significant improvements receive the Earthwards® recognition.
- Product sustainability goals are integrated with business goals.
- Enterprise-wide goals apply across all products and the supply chain.

J&J uses the Earthwards® approach to make their products greener and uses life-cycle reviews to focus on the seven key areas for making individual product improvements (materials, waste, water, packaging, energy, innovation, and social). Environmental advances are also initiated by corporate-wide goals to green all products and the supply chain.

PHILIPS

Headquartered in Holland with 2022 sales of 18.783 billion euros, Philips has a robust greener product program, with a good portion of sales representing more sustainable products. Having three main divisions, Lighting (household and industrial), Consumer Lifestyle (televisions, computers, vacuum cleaners), and HealthCare (CT scanners, ultrasound & diagnostic equipment), makes it challenging to have a unified greener product program. A good deal can be ascertained by studying this European-based conglomerates' approach to greener product design.

In 2020, Philips released statements on "demystifying EcoDesign," as the program is designed to review all parts of the product development process. It is found that approximately **80% of a product's environmental impact can be mitigated in the design phase**. By applying the Life Cycle Analysis, Philips focuses on the following areas:

- Energy
- Packaging
- Hazardous Substances
- Materials
- Circularity

EcoDesigned products that meet one or more of the **"green focal areas"** are indicated as "green products." These products can be seen on the Philips website, indicated as "green products" (Demystifying EcoDesign 2020).

The company made record-breaking goals for Philips' green product sales in 2020 with **73.2% of all sales coming from their "green products,"** which is a

3.8% increase from 2019. In addition, investment in eco-innovation was set at 280 million euros in 2020 (Philips Achieved 2021); in addition to the greener products design program, sustainability goals have been set regularly since the 1990s. The newest sustainability goals have objectives, which apply to all their operations.

PHILIP'S 2025 ENVIRONMENTAL FRAMEWORK

- Maintain carbon neutrality and use 75% renewable energy in operations.
- Generate 25% of our revenue from circular products, services and solutions, offer a trade-in on all professional medical equipment, and take care of responsible repurposing.
- Embed circular practices at our sites and put zero waste to landfill.
- All new product introductions will fulfil EcoDesign requirements, with 'Eco-Heroes' accounting for 25% of revenues.
- Work with suppliers to reduce the environmental footprint of the supply chain in line with a 1.5°C global warming scenario.
- Engage with our stakeholders and other companies to drive sustainability efforts addressing the United Nations Sustainable Development Goals.

(Our Purpose & ESG Commitments 2023)

In order to make the most significant improvements, Philips has identified the key life-cycle aspects for each of their product categories: Health Care—reducing energy consumption, weight, and dose; Consumer Lifestyle—focusing on energy efficiency and closing material loops (e.g., increasing materials recycling), and Lighting—energy efficiency. Examples of the product improvements will help us understand how these goals and strategies are deployed. In the health-care area, Philips launched Ambition 1.5T MR, the company's first ever helium free MRI system. The technology is a massive win for the company and the health-care industry as helium is within limited supply. Over 200 Ambition MR Systems were installed worldwide, saving 306,065 liters of helium (MR Ingenia Ambition 1.5T 2023).

One of Philips newest ventures in the lighting industry was the installation of Philips LED lighting system in the Allianz Arena in Germany. Philips states that this is Germany's first and Europe's largest stadium that features energy-efficient lighting systems. This system **saves about 60% on electricity** and about 362 tons of CO_2 per year. Because of Philips' cloud-based ActiveSite platform, the stadium has lower maintenance and operating costs as well. As we can see, their "greener products" improvements are easily communicated to customers because of their obvious improved environmental performance (Allianz Arena 2023).

PHILIPS GREENER PRODUCT PROGRAM HIGHLIGHTS

- Six Green Focal Areas to make products greener: energy, packaging, hazardous substances, materials and circularity.
- Sustainability goals focus on developing greener products.
- Set enterprise-wide sustainability goals.
- Focus on greening their supply chains.

A FOCUS ON SUPPLIERS

Philips goes beyond greening their products and has set up programs to improve their suppliers' sustainability as well. The programs cover sustainability performance, management of regulated substances, conflict minerals, and other responsible sourcing initiatives. The Supplier Sustainability Declaration (SSD) and Regulated Substances List (RSL) are documents that guide a suppliers' sustainability program. After suppliers agree to Philips' terms and conditions, auditors visit their sites to ensure that they are 100% compliant. If the suppliers do not fully comply, then they cannot work with Philips (Philips Supplier Sustainability Performance 2023).

In 2016, Philips launched the Supplier Sustainability Performance (SSP) program which is characterised by the following four methods:

- A systematic approach to improve sustainability of our supply chain.
- Continuous improvement against a set of recognized and global references.
- Collaboration, increased transparency, clear commitments, and ensuring suppliers meet the agreed targets.
- Encouraging suppliers, industry peers and cross-industry peers to adopt our approach.

(Philips Supplier Sustainability Performance 2023)

To create a frame of reference when assessing and classifying suppliers, Philips utilizes their own requirements and international standards addressing the environment as well as other factors. Depending on the supplier's sustainability score on their self-assessments comprising both their supporting evidence and Philip's screenings are grouped in the following categories: Best in Class (BIC), Supplier Sustainability Improvement Plan (SSIP), Do It Yourself (DIY), and Potential Zero Tolerance (PZT). The PZT category represents suppliers who have violated the Philips Sustainability Agreement, where Phillips will take immediate action to mitigate the situation. Based on one's ranking, Figure 4.1 reveals the necessary steps for supplier's to improve their sustainability performance (Philips Supplier Sustainability Performance 2023).

Phillips has a very effective product sustainability program with most products meeting their greener product objectives. The use of focal areas helps R&D and marketers easily understand how to improve and communicate the improvements to customers. A robust supplier management program also helps with driving more sustainable product develop and supplier environmental performance.

SAMSUNG

The electronics industry has been an early adopter in developing greener products. The very nature of their products is a magnet for NGOs and ecologically aware consumers to put pressure on the manufacturers. The use of electricity and its link to CO_2 emissions, the short life cycle of some of their products, and past poor practices regarding disposal motivate electronic manufacturers to focus on developing greener products. Taking a close look at the approach of a successful global electronics manufacturer headquartered in the Asia Pacific region yields some unique insights. Samsung Corporation, based in South Korea, is interesting to evaluate because of the diversity of electronic products it brings to the market. Some of the products include televisions, cell phones, washing machines and dryers, computers, and printers.

Samsung's approach to greener products started in 1995 when they adopted what they call the LCA approach. Environmental assessments have been required for new product development since 2004 when the "**Eco Design Management Process**" was established. Samsung has set goals to continue their plans for the future to apply recycled resin to all plastic parts by 2050, and apply carbon capture technology to semiconductor production sites, expanding the technology across the company and the supply chain by 2040 (Samsung Sustainability Report 2023).

Early on, Samsung set a goal to develop sustainable innovative products that will reduce their products' impacts across the entire life cycle. Their vision was "Providing a Green Experience, Creating a Sustainable Future." The Eco-Design process consisted of two categories: development process and eco-design activity. The Development Process included: concept, plan, execution, and production. The eco-design activity takes into account the company and product targets, evaluation and improvement of the product, and final verification. Products are put into three categories based on the eco-grading scheme: **eco-product, good eco-product**, and **premium eco-product**. Samsung started measuring these parameters and setting goals to develop more energy-efficient electronics and reduce standby power consumption. Achieving their objectives will not only benefit the environment but also reduce operational costs for customers. They also have adopted a circular resource management system to take into consideration the total environmental aspects in the product design. Every eco-product goes through a rigorous LCA of the following stages: supplier purchases, design and production, distribution logistics and packaging, use and reuse of the product (Samsung Sustainability Report 2023).

A significant focus area for electronics is minimizing the use of the hazardous ingredients; heavy metals and brominated flame-retardants. Samsung implemented a supplier program to address this issue called the Eco-Partner

Certification Scheme in 2004. The company requires the **Eco-Partner Certification** as well as compliance with ISO14000 and OHSAS18000 certifications for all suppliers of products and parts. The object of this initiative is to reduce the use of hazardous parts and raw materials and assist suppliers in setting up environmental quality systems. Suppliers are required to submit raw material data collected along with written evidence that confirms the safety of the specified materials. Samsung then visits the manufacturing locations to verify the information and to provide the Eco-Partner Certification (Samsung Sustainability Report 2023).

Further examples of product improvements are demonstrated through Samsung's partnership with Patagonia and Canadian non-profit marine protection research institution, OceanWise, to create the new Less MicroFiber Filter, an external filter for a washing machine. As clothes are made of synthetic materials, microplastics are constantly generated when worn, washed, and dried. Microplastics are 5 mm of plastic particles (the size of a grain of rice) that pollute waters and oceans impacting ecosystems and biodiversity. Microplastics are found to be too small to filter out of cutting-edge technology. Philip's technology reduces the generation of microplastics by 54% (Samsung Launches Less Micro Fiber Filter 2023). This is a good example of cross industry collaboration and innovation.

SAMSUNG'S ECO-PRODUCT INITIATIVES

- Use the Eco design management process to make greener products.
- LCA is a guiding principle for product improvement.
- Focus on End-of-Life management for electronic products.
- Suppliers of products and parts are to certify to Samsung Eco Partner Standard.

(Samsung Electronics Announces 2022)

Other examples of eco-innovative products include the new Galaxy Z Series mobile phone which utilizes repurposed abandoned fishing nets helping to minimize the impacts of plastic pollution as 64,000 tons of fishing nets are abandoned in the ocean each year. Samsung additionally incorporated post-consumer materials and bio-based resin, where **more than 90% of the Galaxy products are made with recycled materials** (Samsung Repurposes Discarded Fishing Nets 2022).

END-OF-LIFE MANAGEMENT

End-of-life issues are important to consider for any product stewardship program. This is especially true for electronics. Poor disposal practices resulting in environmental and human health problems have been well publicized. Keeping this in mind, Samsung has voluntarily initiated take-back programs for waste electronics. They set up Korea's first waste electronic product recovery and recycling system in 1995. The program is intended to prevent the illegal disposal of electronics through incineration and landfilling and to encourage the recovery of valuable materials.

Expanding on this concept, global take-back programs have been put in place in countries where electronics recycling is not mandatory, like the United States. Samsung collected 1.2 billion pounds of e-waste collected until 2023. The company's goal is to obtain 10 million metric tons of e-waste by 2030 (Responsible Recycling 2023). Samsung has emphasized bringing more eco-products to the market. Through the use of LCAs, efforts have been focused on minimizing the most important impact areas, energy use, and standby mode. Improvements are further encouraged with new product design checklists and guides.

Samsung is a good example of an electronic company that has a long history of more sustainable product development. Having a well-established program that generates products with environmental improvements like energy efficiency and use of post-consumer materials and addressing electronic product end of life are all key focus areas for them.

APPLE INC.

One of the most innovative companies in the world is the electronics manufacturer Apple, the creator of the Macintosh computer, the worlds' thinnest laptop, the extremely popular iPhone, the functional and easy-to-carry iPad, and one of their newest technologies, the Apple Watch. Their innovation has been applied to their sustainability program as well.

Apple had been criticized by environmental groups for not being proactive enough on making their products greener; in 2007, Greenpeace ranked Apple last among major electronics manufacturers on environmental issues. They claimed that the use of hazardous materials like PVC and toxic flame retardants was lagging compared to similar electronic product manufacturers. Since the generation of this report, there has been a concerted effort to make significant product improvements. Eventually, Greenpeace acknowledged that Apple was moving in the right direction. They concluded that their campaign highlighted Apple's shortcomings to their customers, which in turn caused positive changes in their products. Further, they believed that their public reporting of Apple's performance forced Apple to become a green leader (Greenpeace 2023). Regardless of the reasons why Apple has made strides to bring greener products to market, they do present their progress, as you would expect, in an innovative way. Apple has set three focus areas to make the most impact possible. Their first priority is to reduce Apple's impact on climate change through use of renewable energy sources and energy-efficient products and facilities. Their next priority is to conserve resources, and their third priority is to practice smarter chemistry and use safer materials in all products and processes.

Use of Life-Cycle Assessment

Apple uses LCA to measure its environmental footprint. The key indicator used for determining their footprint is greenhouse gas emissions. For 2022, Apple estimates that it emitted a total net carbon footprint of 20.3 million metric tons of greenhouse gas emissions. Emissions are calculated from corporate and product life-cycle emissions. Their analysis determined that a majority of their emissions come from product manufacturing (75%) and customer use (25%) (Apple Environmental Progress 2023). Therefore, to improve their footprint, they decided to focus on sourcing lower carbon materials, partnering with suppliers to add clean energy to their facilities and produce clean energy at Apple offices, retail stores, and data centers globally, as well as adjust recycling and shipping strategies. To address the use of materials in their products, a DfE concept, dematerialization, was employed very successfully. Looking at their product lines, it's evident that design engineers have fostered the innovation of thinner, smaller products that deliver the same or more benefits as competitors. The reduction of raw materials used in electronics yields significant environmental improvements. Consider the impact benefits from fewer metals having to be mined and transported to manufacturing sites, lighter products which require less energy and chemicals to process and less transportation emissions are generated since more packages can fit on trucks or planes during shipping. An example of circularity combined with enhanced performance is the iPhone 14 and the iPhone 14 Plus. Representing pioneering chemistry R&D, the phones' cameras wires are composed of 100% recycled gold. The phone series consists of a Ceramic Shield and water and dust resistance that help with longevity and durability (iPhone 14 Environmental Report 2022).

APPLE'S ECO-DESIGN FOCUS AREAS

- Use less material.
- Ship with smaller packaging.
- Be free of toxic substances.
- Be as energy efficient and recyclable as possible.

REMOVING TOXIC MATERIALS

Raw materials needed for electronic products contain toxic materials. Since Apple received a good deal of criticism for not being proactive in removing toxic materials from their products, they decided to be more public about their accomplishments and future goals. Apple has reached their goal to lead their industry in reducing or eliminating environmentally harmful substances such as toxic metals, brominated flame retardants (BFRs), phthalates, and polyvinyl chloride (PVC). A typical cathode ray tube (CRT) contains an amazing amount of lead (approximately 3 pounds or 1.36 kg). Apple became the first company in the computer industry to eliminate CRTs back in 2006. At the time, all the leading manufacturers of PCs were still using CRTs (Apple Inc. 2010). Apple claims that they are far ahead of their competitors in reducing toxic materials. Every product is free of BFRs, elemental bromine, beryllium-free and chlorine and every display has mercury- and arsenic-free glass. Continuing their leadership in toxic material removal, Apple completely removed all PVC from their power cords and headphone cables. This four-year process eventually led to the use of safer materials. Apple found the perfect blend of durability, safety, and environmental performance. They have replaced all PVC with non-chlorinated and non-brominated thermoplastic elastomers.

PACKAGING

Another objective to improve product environmental performance is to minimize product packaging. Apple's efforts have a double benefit—less packaging materials and reduced greenhouse gas emissions during product transportation. The company is close to achieving their goal of eliminating all plastics from their packaging by 2025, where they have a remaining 4%. With such improvement, over 1,100 metric tons of plastic and over 2,400 metric tons of CO_2 emissions were avoided in 2022 (Apple Environmental Progress 2023).

ENERGY EFFICIENCY

Focusing on designing more energy-efficient products is one of the most critical aspects of eco-design. Apple's latest generation of the Pro Chips that can be found in the MacBook Pro M2 Pro and M2 Max as well as the Apple TV 4K utilizes 30% less power than previous generations. The A15 Bionic Chip eliminates the need of an internal fan, creating a more compact design and reducing the carbon footprint by 25% (Apple Environmental Progress 2023). Having third-party standards to verify claims is helpful to build credibility. Energy-efficiency claims have been confirmed by meeting or dramatically exceeding the US EPA's Energy Star criteria. "In 2022, **all eligible Apple Products received Energy Star rating for superior energy efficiency**." Making products that last longer is a design-for-environment principal. A longer lasting product reduces the resources needed to make new ones (Apple Environmental Progress 2023).

Recycling Initiatives

No matter how long a product is designed to last, it has an end. To lower their carbon footprint, Apple has placed a priority in recycling products in the region where they're collected to reduce harmful carbon emissions. Electronic products have considerable concerns because of its metallic components and flame retardants. Many electronic manufacturers have established end-of-life management programs to address this concern. **Apple has established recycling initiatives in 99% of the countries where their products are sold.** In 2022, through customer and employee programs, Apple recycled more than 40,000 metric tons of electronic scrap globally (Apple Environmental Progress 2023).

Part of this initiative was to ensure that recycling partners are acting responsibly; therefore, auditing programs and guidelines are critical. Apple does not permit firms to dispose of electronic waste in solid waste landfills nor permits them to be incinerated; this is not always a guarantee in developing nations. Enabling the take-back and recycling of products helps to bolster a company's green image and is a strong statement to customers, showing that environmental issues are front of mind when a product needs to be disposed of. To address this issue further, the Material Recovery Lab (MRL) has created an automated approached to material recovery from robots, Daisy, Dave, and Taz. Daisy, the disassembly robot, takes apart 23 different types of iPhones, ranging from the iPhone 5 and iPhone 12. Daisy can recover 2,000 metric tons of mined rock from one metric ton of iPhone (Apple Environmental Progress 2023).

Product Environmental Report

Several companies publish product-specific environmental reports, which I think is really a good idea because it's a single source of truth for all the environmental information a customer or stakeholder would want for a product. Apple has posted several reports on their website. Since their latest innovative product, the Apple Watch, a review of how Apple evaluates this product will help in understanding their approach to product greening and demonstrate a best practice for communicating a product's environmental footprint.

The categories that are addressed for each product include:

1. Climate change
2. Energy efficiency
3. Material efficiency
4. Packaging
5. Smarter chemistry
6. Recovery

The report details the Apple Watch Ultra environmental impacts with graphs and tables. The life-cycle greenhouse gas emissions are estimated at 56-kg CO_2e. The emissions from the greatest to least are reported as 77% production, 11% customer use, 11% transportation, and <1% recycling. The company is shifting to have all Apple Watch Ultra final assembly supplier sites to utilize 100% renewable energy to help reduce its carbon footprint (Apple Watch Ultra Environmental Report 2022).

Apple strives to use recyclable materials like aluminum and glass. They are leading the industry in ultra-compact product and packaging designs which reduce the overall material footprint of a product. The Apple Watch Ultra uses 100% recycled rare earth elements, tungsten, tin, gold, and up-cycled plastic water bottles (Apple Watch Environmental Report 2022). Packaging is addressed by the type and weight of material for the retail box and shipping box, for example, virgin wood fiber and minimal amounts of plastics. The Apple Watch Ultra retail box contains at least 38% recycled content and is the first apple watch to contain 100% recycled gold in the plating of circuit boards (Apple Watch Ultra Environmental Report 2022).

To show progress toward the use of restricted substances, the Apple Watch Ultra complies with the European Directive on the Restriction of the Use of Certain Hazardous Substances in Electrical and Electronic Equipment (RoHS Directive). This Directive covers lead, mercury, cadmium, hexavalent chromium, and polybrominated biphenyl (PBB) and polybrominated diphenyl ethers (PBDE) brominated flame retardants (BFRs). Apple products also comply with the European Regulation on Registration, Evaluation, Authorization and Restriction of Chemicals (REACH). Apple reports that they have reduced the following hazardous ingredients even further than regulatory requirements: mercury, BFRs, PVC, and beryllium. Recycling is reported in a general manner in the product report. Use of less raw materials and inclusion of recyclable materials in the product design are emphasized; the result is less waste at end of life, plus maximum recyclability. In addition to the very detailed environmental report, a status report is maintained to highlight some key features.

APPLE WATCH ULTRA ENVIRONMENTAL STATUS REPORT

Apple Watch Ultra is designed with the following features to reduce environmental impact:

- Mercury-free
- Brominated flame retardant-free
- PVC-free, beryllium-free
- Complies with European REACH regulation on nickel
- Retail packaging contains at least 38% recycled content.

Apple is a leader in innovative products, and they have taken an innovative approach to product stewardship. One of their most significant accomplishments is the light sizing of electronic devices. They have been a leader in making some of the thinnest laptops and smartphones in the electronic industry. This initiative alone has gone a long way to advance greener products. Strides have also been made in reducing toxic materials and designing devices for longevity. Apple has some of the longest lasting batteries in the electronics industry. Providing end-of-life recycling solutions for customers addresses an important environmental concern for electronic products. The use of **environmental reports on individual products** is an industry leading practice that helps to catalogue the environmental improvements on a product level.

SEVENTH GENERATION

Seventh Generation is a company that from its very beginning was based on bringing greener products to market. Sustainability is the very essence of the company. In their mission, they state an Iroquois Indian law, which shaped the name of the company, that they must **consider the impact of their decisions on the next seven generations**. You can't have a more solid foundation for a company to base their greener product program on. So, it makes sense to evaluate the approach of Seventh Generation to understand how a company with green at its very roots develops products (Seventh Generation: More Than a Name 2022). Seventh Generation makes various home use products such as botanical disinfectants, dishwashing detergents and soaps, hand soap, laundry detergents, surface cleaners, household paper and supplies, baby diapers and wipes, baby laundry, organic tampons and pads. Their product developers, while focusing on effectiveness, are to consider economic, environmental, and health factors. A product scorecard is used to assist in developing greener products.

SETTING SUSTAINABILITY GOALS

The use of sustainability goals to drive product improvements is part of Seventh Generation's strategy. Objectives have been set to reduce impacts like greenhouse gas emissions and increase the use of plant-derived ingredients. One unique approach is that they have set an internal price on carbon at $12/ton. They hope this will encourage purchasing groups to buy materials that have lower carbon associated with it.

2025 SUSTAINABILITY GOALS

Decrease Our Carbon Footprint
- Reduce Scopes 1 through 3 (indirect and direct emissions) inclusive of consumer use—by 50% compared to 2012.
- The water cycle will not be contaminated during a product's life cycle.

Sustainable Sourcing
- 100% of materials and ingredients will be bio-based or recycled.
- 30% of palm kernel oil will be replaced by non-land use, bio-based oil.

Zero Waste
- 100% of Seventh Generation's packaging will be reusable and reused or recyclable and recycled.

(Seventh Generation Journey 2025 Goals 2023)

PACKAGING

One of the most visible aspects of a consumer product companies' sustainability is packaging. Realizing this, Seventh Generation has made a conscious effort to place greener packaging on store shelves. Significant progress was made on the eco-friendliness of the type of material used and use of recycled content. Setting high expectations for your organization is a necessary step in developing greener products. This includes setting stretch goals such as striving for 100% post consumer recycled (PCR) content in bottles before anyone else in your category. Leading isn't easy; "failed attempts" due to PCR rigidness created problems. Through their partnerships with suppliers, progress was made and now hand dishwashing liquid, fabric softener, and bleach containers use 97% PCR (The Race to Zero Waste Seventh Generation 2019). The company continuously looks for different methods in shifting away from plastic with their new Seventh Generation **Zero Plastic Products Line**. The Zero Plastics line consists of steel canisters **that are considered to be the world's most recycled and recyclable material** (Zero Plastic Line Seventh Generation 2021). Post-consumer content packaging has increased over time, and Seventh Generation has also emphasized the use of renewable materials. Packaging goals for 2025 strive for all packaging to be bio-based or recycled, biodegradable, or recyclable. **The materials that are used for plastic packaging follow this hierarchy: post-consumer recycled plastic, virgin plant-based plastic, and virgin petroleum-based plastic** (Corporate Conscious Update Seventh Generation 2020).

MATERIALS

Having a natural products line requires a greater focus on the ingredients that go into a product. Therefore, relentless effort to improve your formulas is critical to maintaining your customers' trust. An example of the continual greening of products is the Early Dose Ultra Concentrated Laundry Detergent. The innovation uses 60% less plastic, helping to reduce its carbon footprint, and is 75% lighter compared to the 90-fl-oz bottle (EarlyDose Ultra Concentrated Laundry Detergent 2023). Future directions for greening products are guided by Seventh Generation's new sustainability goals: all products and packaging create **zero waste and are biodegradable or recyclable by 2025**; all **ingredients and materials are bio-based or recyclable by 2025**. Another ingredient activity that demonstrates attention to detail is the commitment to sustainably sourcing materials containing or derived from palm oil. Palm oil has become a significant issue because it is sourced in regions where portions of rainforests have been cut down to establish palm groves. To address this issue, Seventh Generation joined the Roundtable for Sustainable Palm Oil (RSPO) in 2008. Through the program, the company continues to purchase and use RSPO-certified products. A goal was established to have 30% of palm kernel oil replaced by bio-based oil in 2025 (Corporate Conscious Report Seventh Generation 2021).

Unique ingredients have been used to offer customers products with more natural materials. Seventh Generation earned its certification from the USDA BioPreferred Program, where all products are 95% bio-based or higher (Bio-Based Plant Based Seventh Generation 2018). New disinfectant cleaners were developed using thymol, a component of thyme oil derived from the garden herb thyme. Seventh Generation claims that these cleaners kill 99.99% of germs. A review of the ingredient list from their Disinfecting Multi-Surface Cleaner demonstrates a commitment to natural ingredients. It reads like a label on a salad dressing bottle.

DISINFECTING MULTI-SURFACE CLEANER INGREDIENT LIST

Thymol 0.05% (component of thyme oil), aqua (water) sodium lauryl sulfate (palm kernel or coconut-derived cleaning agent), sodium citrate (cornstarch-derived water softeners), citric acid, and copper sulfate pentahydrate (bluestone) (mineral-derived water mineralizer), fragrances and essential oils: Cymbopogon nardus (citronella) oil, Cymbopogon schoenanthus (lemongrass) oil, citral (fruit peel oil) (Disinfecting Multi-Surface Cleaner Seventh Generation 2023).

Transparency is another key aspect to Seventh Generation's commitment to greening their product ingredients. They commit to disclose all ingredients in their products and do so on their website and on product labels. A visit to their website will enable the viewer to see a list of everything that is in a Seventh-Generation product through their collaboration with SmartLabel, a platform intended for consumers to learn more about ingredients. The list of ingredients above for their Multi-Surface Disinfectant Cleaner is from their website. There is also a glossary that explains terms used to describe ingredients used like synthetic, plant derived, plant modified, etc. as well as its function for the product. This is a best practice because it makes it easy for customers to understand what the purposes of ingredients are and how they are being derived.

KEY ASPECTS TO SEVENTH GENERATION'S PROGRAM

- The company is based on having greener products: They consider the impact of their decisions on the next seven generations!
- Big emphasis on natural ingredients & incorporation of PCR in packaging and recyclability.
- Set sustainability goals to encourage greening products.
- Commitment to transparency of all product ingredients.
- Focus on removing plastics from packaging.

Seventh Generation is a company whose very essence is based on bringing greener products to the marketplace. It is obvious that they are fully committed to this concept when reading their sustainability report. There is a big emphasis on natural ingredients, removal of petroleum plastics, and the use of PCR in their packaging. A further focus on sourcing ingredients in a sustainable manner and a transparency commitment to listing all ingredients for each product sold rounds out the key elements to their product greening strategies.

METHOD

You need not go any further than the company tagline "people against dirty" to understand Method's commitment to sustainable products. The company was formed in 2001 with the goal of providing natural cleaning products. Looking at their interactive website, you know you're dealing with a cutting-edge, innovative company. Method makes shampoos, lotions, deodorant, hand cleaners, body wash, laundry detergent, and household cleaners. Their goal is making products "future friendly" (Method Products Packaging 2023). Of all the companies evaluated in this book, Method has one of the most comprehensive approaches to greener product design. According to company statements, they believe that business can be an agent for positive social and environmental change. Method was one of the first companies in 2007 to become a B corporation, a comprehensive third-party sustainability certification. A list of criteria is used when designing products for "true sustainability." Method rightly states that focus cannot be on just one element like carbon footprint or safety; new products require a holistic evaluation. Method explains that only focusing on one aspect to green a product may result in less than optimum results. For example, natural should not be the only dimension that is focused on for a product. Natural ingredients are important, but some natural ingredients can be toxic. Method uses natural as the place to begin and also evaluates all other aspects of a product to insure it's making the most sustainable decisions possible.

PACKAGING

Sustainable product assessments must also include consideration of the product packaging. An important consideration for packaging is the recyclability of the materials being used, or as Method states, using "bottles made from bottles." The use of PCR is paramount to designing sustainable packaging according to Method; they state that all PET plastic bottles use 100% recycled plastic, which have an 70% lower carbon footprint than using products that are made of virgin plastic. Their HDPE bottles consist of about 25% of PCR in toilet cleansers to 50% PCR in their 8× laundry detergent. They also strive to reduce the mass of the plastic in each bottle, which is evident by their marketing of **refill pouches** which **save 78–82% water, energy, and plastics compared to a bottle** (Method Products Packaging 2023). Method also launched recyclable and refillable aluminum soap bottles that provide not only environmental benefits but also a sleek and clean aesthetic. The new line will allow for consumers to save 96% of plastics as they will refill their bottles with the **concentrates** (Refill + Reuse Method 2023).

CRADLE-TO-CRADLE DESIGN

Method says that they follow the precautionary principle regarding ingredients. A way to demonstrate this commitment is embracing the Cradle-to-Cradle (C2C) design theory. This theory was developed by McDonough Braungart Design Chemistry (MBDC), a sustainability-consulting firm. This firm also issues certifications for products that follow their **Cradle-to-Cradle® Framework**. The certification is issued on a product-by-product basis and covers material health, material reutilization, water, energy, and social fairness (MBDC 2016). C2C design encourages following natural cycles, or as MBDC says, "nature's biological metabolism." The main idea behind C2C is to design products and materials with life cycles that are safe for human health and the environment and that can be reused in a closed-loop fashion (MBDC 2016).

KEY BUSINESS BENEFITS OF CRADLE TO CRADLE

- Evaluates products
- Reduces risks
- Reduces costs
- Third-party independent evaluation and certification

The C2C design framework has been included in all the products that Method designs. In 2016, they had over 100+ products that were C2C certified, one of the highest numbers of any company. Products such as hand wash, dish detergent, laundry detergent, and body wash are certified. It is Method's goal to have their entire product line achieve certification. The C2C criteria are very comprehensive. There are five categories in which a product is evaluated. In each area, there are criteria set that lead to achievement of silver, gold, and platinum levels.

C2C CATEGORIES OF CRITERIA FOR CERTIFICATION

- Material health
- Material reutilization
- Water use
- Energy use
- Social fairness

The C2C product program assesses the environmental effects, hazards, and risks of the product that is being evaluated by looking at a complete materials list of the product and the product packaging. As an example of how products are evaluated, the Material Reutilization category "rewards products that contain recycled or renewable materials." Manufacturers are expected to develop and implement a strategy to close the loop on the product at the end of its useful life, at the highest levels of certification (MBDC 2016).

Formulation Approach
The following five steps are used when developing products:

- *The precautionary principle*: If there's a chance an ingredient isn't safe, we don't use it.
- *The dirty ingredient list*: Ingredients that many others use, but we don't.
- *The highest standard*: Creating C2C products.
- *Comprehensive, third-party assessment*: Ingredient reviews are completed by independent researchers using scientifically peer-reviewed materials.
- *Smart science*: Green chemistry techniques ensure effective and safe formulas.
- *External validation*: C2C certifications.

Having a robust process is critical to bringing greener products to market. This includes having principles and procedures that product developers can use. As indicated above, **Method uses a comprehensive evaluation process to bring greener products to market**, giving product developer's guidance about which materials to steer away from is an important element. A list of undesirable or "**dirty ingredients**" is maintained which formulators must avoid. The list includes materials such as chlorine bleach, triclosan, EDTA, phosphates, 2-butoxyethanol, phthalates, paraben, ammonia, and MEA (ethanolamine) to further bolster trust in their brand, each product page on their website lists every ingredient and fragrance used. Other companies have committed to this form of transparency; however, Method goes further and **makes available an evaluation of each ingredient by a third party**. Product labels have a list of the ingredient names and a link to where you can get more information online.

KEY ASPECTS TO METHOD'S PROGRAM

- The company mission is founded on greener products: "People against dirty".
- Holistic greener product design process & guides.
- Commitment to C2C certification.
- Third-party assessment of all products and ingredients.
- All product ingredients are made publicly available.
- Emphasis on the use of PCR in packages.

Method has been a deep green company from its very beginning. All of their products are based on greener product design principals. They consider an extreme amount of details when developing products from the efficacy and toxicity of raw materials, avoiding the most problematic (or dirty) chemicals, to the amount of PCR in packages. All products undergo a third-party review to ensure that they meet the company's goals for greenness. Method has committed to getting all products certified by C2C. All of the product ingredients and fragrances are available on their website. The company believes that they can make a difference with their products and are committed to bringing healthier and environmentally safe products to market.

PROCTER & GAMBLE

Procter & Gamble (P&G) is a company with over 170 years in business and, with $82.006 billion in 2022 sales, is the world's largest consumer packaged goods company. They have iconic brands which are sold globally. Some of their key brands are Tide laundry detergent, Luvs disposable diapers, Gillette razors, Olay beauty products, and Duracell batteries. Being a large multinational company with over $82 billion in sales (2022), it is interesting to see how they have made their sustainability programs and initiatives very public and an obvious enhancement to the company's equity. A big part of their sustainability initiative is their product greening efforts.

THE USE OF GOALS TO IMPROVE PERFORMANCE

P&G received significant publicity when they announced their long-term environmental vision. Of these long-term goals, two of them focus on their products, which demonstrate that they understand that their biggest environmental and social impacts come from manufactured goods.

P&G'S ENVIRONMENTAL VISION & OBJECTIVES

- *Climate*: Net zero ambition by 2040.
- *Waste*: 100% renewable or recycled materials will be used for all products and packaging by 2030.
- *Water*: Restore more water than is consumed by 2030.
- *Nature*: Protect, restore or improve greater than 1.5 million acres of land important for people, biodiversity and P&G.

(Citizenship Report 2022)

Having bold goals is good, but it's necessary **to set interim targets to demonstrate progress** that will put the company on a path to realizing the bigger objectives. P&G has set 2030 and 2040 targets as stepping-stones toward their long-term vision. In their 2022 Citizenship Report, progress has been made on all their long-term objectives, where P&G remains transparent with consumers on their current status. Let's focus on the progress they reported associated with their products; it's a good list of positive actions a company can take to make meaningful improvements to their products. The company has **79% of its products designed to be recycled or reused**. They doubled their use of recycled plastic over the past two years. P&G also focuses to **minimize their use of virgin petroleum plastic in packaging** by 50% by 2030. In their report, they reveal that they have reduced about 8% of virgin petroleum plastic since 2017.

The company also achieved third-party certifications for palm oil (100% RSPO certified), wood pulp (100% third-party certified), and paper packaging. P&G intends to uphold their responsibilities and maintain certifications for the palm oil and wood pulp. In terms of paper packaging, they achieved Forest Stewardship Council (FSC) certification for 68% of their virgin paper packaging (P&G Citizenship Report 2022)

PRODUCT INNOVATION USING A LIFE-CYCLE APPROACH

To reduce product environmental impact and spur product innovation, P&G uses an LCA approach. Looking at a product's full life-cycle impacts, from raw materials to manufacturing and product use, helps identify the most important areas to focus on to make the greatest reduction in environmental impact. Prior to consideration of life-cycle impacts, companies focused primarily on reducing the footprint at the manufacturing facility. Environmental initiatives that were focused on reducing energy, water use, and waste generation all seem like the right thing to do. However, LCA made it clear that the biggest impacts for most companies are the raw materials they use or the use or end-of-life phase of a product. A good example is the application of LCA is Dawn Powerwash, an analysis demonstrates that much environmental impact occurs in the consumer-use phase. Dawn Powerwash "gets dishes done faster" through its specialized formulation to eliminate the use of water where "no soaking, pre-rinse, or re-rinse is needed." When handled properly, the product is said to **reduce the water needed by 50%** and even won the Good Housekeeping 2021 Sustainable Innovation Award (Mohan 2022).

Another example of the use of LCA is the evaluation of using a mop and bucket for cleaning. This enabled product developers to focus on the most important area and generated a significant sustainability innovation. The Swiffer Wet Pad is a more convenient way to clean which allows the consumer to combine a mop, bucket, and cleaning solution. A household that uses this product will save more than 70 gallons of water every year in comparison to a mop and bucket—more convenient and lower environmental impact.

The evaluation of Mr. Clean's Magic Eraser indicated a need for a new packaging design that would be light in weight and sustainable. The company switched to using PET tub designs to a paper based recyclable carton, reducing the use of plastics by 482 tons per year equivalent to the weight of 48 million plastic water bottles. The package design won two Excellence Awards in the 2018 Paperboard Packaging Council carton competition (Mohan 2022).

PROCTOR & GAMBLE'S GREENER PRODUCTS PROGRAM

Long-term Goals
- Improve sustainability efforts among the four pillars: Climate, Waste, Water and Nature.
- Designing products that delight consumers while maximizing the conservation of resources.
- Use LCAs to work on the most important product impacts.

P&G is the largest package good consumer product company in the world. Marketing products that are prominent in most households gives P&G a tremendous opportunity to make big impacts with greener products. Very bold goals have been set demonstrating significant commitment to reducing the impacts of their products. The use of LCA guides product developers to focus on the most important areas needing product improvement, which have led to innovations. P&G sets a high bar for consumer product manufacturers to follow and indicates that there is a strong pull for greener products in the marketplace.

UNILEVER

Another large consumer's product manufacturing company which has made significant sustainability commitments is Unilever. Headquartered in London, Unilever states that **on any given day, 2 billion people use their products**. With sales of 60.07 billion euros (2022), they manufacture a lot of food products, refreshment categories, home cleaning, and personal care products. They have over 400 brands, many of which are household names, such as Lipton, Knorr, Axe, Hellmann's, Heartbrand Ice Creams, Magnum, Rama, Ben & Jerry's, Rexona, Sunsilk, and many more. When you manufacture products used by 2 billion people every day, you have a big opportunity to impact the world.

> We're a company of brands and people with a clear purpose: to make sustainable living commonplace
>
> *(Unilever Planet & Society 2023)*

The company's vision clearly embraces sustainability. To fully understand how this initiative impacts product design, we will evaluate some of the accomplishments and direction Unilever has taken. Unilever set a bold commitment to a long-term sustainability initiative through their Unilever Compass Strategy in 2023. The plan embraces all aspects of sustainability and ties them into sourcing and selling their products. The idea is to increase sales with sustainability in mind "to help us deliver superior performance and drive sustainable and responsible growth" (Unilever Planet & Society 2023).

UNILEVER COMPASS STRATEGY

The Unilever Compass Strategy has three categories of focus:

- Improve the health of the planet.
- Improve people's health, confidence and wellbeing.
- Contribute to a fairer, more socially inclusive world.

(Unilever Compass 2023)

IMPROVING THE HEALTH OF THE PLANET

Unilever focuses on the following three different sectors in regard to the environment: **Climate Action, Protect and Regenerate Nature**, and **Waste-Free World**. Under Climate Action, the company focuses on achieving zero emissions in its operations by 2030 and net zero emissions across its value chain by 2039. To fulfill its objectives, Unilever is "transitioning to renewable energy across our operations, finding new low-carbon ingredients, expanding our plant-based product range, and developing fossil-fuel-free cleaning and laundry products." Their product focus is working towards the implementation of a deforestation-free supply chain of specified materials, the protection and regeneration of 1.5 million hectares of land. The third final category, Waste-Free World, focuses on reducing 50% virgin plastic use by 2025, applying 25% of recycled plastic by 2025, and utilizing 100% reusable, recyclable, or compostable plastic packaging by 2025 (Unilever Compass 2023).

Improve People's Health, Confidence, and Well-Being

Unilever plans to **enhance the livelihoods of millions of people** as they grow their business. There are two key areas: **Positive Nutrition** and **Health and Well-Being**. The focus of positive nutrition is to re-evaluate certain products like ice cream and look for ways to make certain ingredients plant-based and reduce sugar/calorie intake for consumers. The Health and Well-Being area looks for ways to improve the livelihoods of people worldwide and promote the diversity, equity, and inclusion (DEI) principles for 1 billion people per year by 2030 (Unilever Compass 2023).

Contribute to a Fairer, More Socially Inclusive World

Unilever takes initiatives on the following social issues: DEI, Raise Living Standards, and the Future of Work. For DEI, Unilever looks to create a fostering environment eliminating bias and discrimination from the workplace. Working on employee engagement, Unilever works towards compensating its employees with a fair and viable wage. To help build a better future, the company plans to work with 10 million young adults, equipping them with essential work-related skills by 2030 (Unilever Compass 2023).

Reducing Emissions

So how does a company reduce emissions not under their control? Unilever is using its own platform to encourage a "system-wide change" that will help to **reduce greenhouse gas emissions in consumer homes**. The company teamed up with global campaigns that focus on bringing renewable energy, like Renewable Electricity 100 (RE100) and Powering Past Coal Alliance. Since 2010, Unilever saw a 19% reduction in greenhouse gas emissions per consumer.

The company is committed to bringing its consumers lower carbon footprint products through its **Clean Future Strategy** ("focus on renewable and recyclable carbon ingredients"), **Future Foods Strategy** ("focus on plant-based foods"), and **Positive Beauty Strategy** ("focus on sustainable sourcing, deforestation-free palm oil, and nature-based solutions"). With an allocated Climate & Nature Fund of 1 billion euros, all brands will undergo a greening process (Reducing Emissions Unilever 2023). An important environmental aspect for a company that sells food products is responsible sourcing initiatives. These initiatives are becoming a greater area of focus for companies' greener product development programs. Let's take a deep-dive look at the plan, which I believe is one of the most comprehensive and an industry best practice.

Sustainable Sourcing

Working with suppliers can make a much bigger impact than focusing on a company's own manufacturing operations. Many of Unilever's products depend on agriculturally derived raw materials. Some developing countries have depleted important biological areas like tropical rain forests to produce agricultural products. NGOs have urged focusing on sustainable production methods to combat this problem.

> Unilever has made it a goal to achieve 100% of their agricultural products from raw materials that are sustainably sourced—a leading practice.

Unilever has made it a goal to achieve 100% of their agricultural products from raw materials that are sustainably sourced. They created a program called the **Unilever Sustainable Agriculture Code**, a 72-page document of guidelines on how to sustainably farm key crops, which **all their suppliers must comply with**. In addition to this Code, there is a **"Responsible Sourcing Policy"** which also has mandatory requirements for suppliers that include regulatory/legal compliance, fair employment practices, protection of indigenous people and communities' land rights, and sustainable operations. The company is aware of its position as a leading purchaser of palm oils and teas (depleting agricultural raw materials) worldwide. The company is committed to sustainably sourcing these two key commodities, proven through their efforts as Unilever sourced 81% of key agricultural materials sustainably in 2022 (Sustainable & Regenerative Sourcing Unilever 2023).

Palm Oil

Perhaps the poster child for needing to be responsible for sourcing agricultural commodities is palm oil. Of all the agricultural raw materials, palm is most noted for deforestation, forced labor, habitat degradation, and indigenous people rights abuses in the countries where it is produced, primarily because land and forests are cleared to make way for the development of palm oil plantations. As the awareness of how unsustainable farming palm oil is became increasingly apparent, the more businesses realized that they needed to take steps to address this issue. Unilever was one of the first companies to make a goal back in 2009 to commit to **sustainably sourcing 100% of their palm oil.** The company reached this goal as of 2015, all palm oil has been sourced sustainably. They were able to reach their goal three years ahead of their plan due to the development of GreenPalm certificates. Growers can earn GreenPalm certificates from the RSPO. Unilever helped form the RSPO—an initiative working with the NGO, World Wildlife Fund (WWF), to develop standards for sustainably producing palm oil. In 2022, Unilever sustainably sourced 94% of their core palm oil, where 86% comes from physically certified sources and 8% from RSPO independent smallholder credits (Sustainable & Deforestation-Free Palm Oil Unilever 2023). To take their program to higher levels they developed a partnership with the WRI, a "traceability and risk verification system on the ground with WRI's Global Forest Watch Platform" is being developed. This is an online monitoring and alert that uses satellite technology, open data, and crowd sourcing to guarantee access to timely and reliable information about forests.

Unilever is very transparent about their plans and pledges; an important way to gain consumers trust. On their website, they make the following commitment:

A deforestation-free supply chain in palm oil and other agricultural raw materials by 2023. Unilever was the first consumer goods company to disclose its palm oil suppliers/mills list, even stating what is and isn't RSPO certified. The company also goes as far as to reveal grievances and issues surrounding palm oil publicly, showcasing that there is much work that needs to be done; "We are therefore applying cutting-edge technology solutions to get a clearer view of our supply chain, mapping smallholder's communities, expanding our direct sourcing programme with independent mills, and investing in landscape approaches to drive impact at speed and scale. By continuing to work with our stakeholders, we can eliminate deforestation, protect peatlands, support smallholders, and drive positive impacts for people and local communities" (Sustainable & Deforestation-Free Palm Oil Unilever 2023). These commitments put Unilever in a leadership position on this issue with some pretty big goals.

WATER

Many of Unilever's products require the use of water, so a goal was set to implement water stewardship programs in 100 water stressed locations. Additionally, the company is creating "water-smart products" to help reduce the water consumed by its customers. As an example of how this is put in practice, the Dutch branded product, Robijn dry-wash-spray, is utilized to refresh clothes between washes, requiring less laundry and saving water. Another product, Rin detergent bar, **only uses half the water needed for rinsing**, creating a beneficial outcome for consumers (Water Stewardship 2023).

Waste and Packaging

Unilever is working towards making their packaging 100% recyclable, reusable, and compostable. Part of achieving this goal is the commitment to developing a circular economy through the Fair Circularity Initiative, a program committed to respecting worker's rights within the waste sector (Rethinking Plastic Packaging 2023).

Future Foods

Food disparities are persistent in today's global society, where 2 billion people struggle with obesity and another billion endure hunger. Amplifying the matter, the food industry contributes to more than 20% of all greenhouse gases and a third of all food produced globally is thrown away. Unilever's President of Nutrition, Hanneke Faber, expresses, "As one of the world's largest food companies, we have a critical role to play in helping to transform the global food system. It's not up to us to decide for people what they want to eat, but it is up to us to make healthier and plant-based options accessible to all." As a result, the company launched Future Foods, a campaign to provide consumers healthier diets while simultaneously reducing their environmental impacts (The World's Food System Unilever 2023). The objectives of Future Foods are as follows:

- An annual €1.5 billion sales target by 2025 from plant-based products in categories whose products are traditionally using animal-derived ingredients.
- Halve food waste in our direct operations from factory to shelf by 2025.

- Double the number of products that deliver positive nutrition by 2025.
- Continue to lower calories, salt and sugar across all our products.

(The World's Food System Unilever 2023)

In creating greener products through a health lens, Unilever has reduced the sugar content in Lipton Iced Tea by 30% across Europe and introduced a limit of 110 calories in children's ice cream. Additionally, the company teamed up with Wageningan University & Research (NL) to create meat and dairy alternatives, with plant-based foods comprising a third of their portfolio. Unilever is looking to reduce their use of plastics among the food industry. As an example, Hellman's launched their 100% recycled and recyclable containers in the United States and Canada, saving approximately 14,000 tons of virgin plastic per year (The World's Food System Unilever 2023).

UNILEVER'S GREENER PRODUCT PROGRAM

- Developed very aggressive long-term goals through their Compass Strategy.
- Strong CEO support for their program.
- Significant commitment to sustainable sourcing.
- Commitment to educating consumers to reduce their environmental impacts.
- Partner with suppliers to drive sustainable innovations.

Unilever has a large impact on the world's resources, purchasing significant amounts of raw materials and selling billions of consumer products. Having a company vision that embraces sustainability enables the development of greener products. The use of lifecycle thinking facilitates the focus on reducing environmental impact of the product at key life-cycle stages where it makes the most sense. Unilever has developed education programs to help consumers reduce their environmental impacts. Being reliant on agricultural commodities, a very comprehensive sustainable sourcing program was initiated to partner with suppliers to reduce impact at the very beginning of their products' life cycle.

REFORMATION FASHION

In 2021, the fashion apparel retail industry was estimated to have generated a revenue of $1.7 trillion dollars worldwide (Global Fashion Industry 2023). This industry is one of largest in the world and is also one of the most polluting. One of the reasons why the fashion industry is detrimental is due to the extensive supply chain involved in making a garment. There are many steps necessary to bring a product to market and all of them have significant impacts: production, raw material, textile manufacture, clothing construction, shipping, retail, use, and ultimately disposal of the garment. The carbon footprint in this industry is extremely large. In order to get one article of clothing to the consumer in a developed country, it typically travels thousands of miles since a lot of the manufacturing is performed far from the stores where the products are sold in. In addition to carbon emissions, process steps are quite polluting. Consider the pesticides used for cotton, plus wastewater, toxic dyes, and emissions from manufacturing synthetic textiles, and at the end of life of a garment, clothing that needs to be disposed of, just to name a few. The rise of fast fashion and over consumption has only increased sustainability issues for this industry. Apparel companies are starting to realize the effects of the fashion industry on the environment and are starting to take initiative to address these concerns.

Starting as a small Los Angeles vintage shop in 2009, Reformation, a women's retail brand, has its focus on bringing sustainable fashion to everyone. The company delivers sustainable and high-quality clothing and the company continues to grow rapidly, having sales doubled in four years to hit $300 million (Rockeman 2023). Although their factory is based in Los Angeles, Reformation collaborates with people from around the world ensuring equitable and safe working environments (Reformation About Us 2023).

SUSTAINABILITY INITIATIVES

Reformation has developed a sustainability scorecard with the support of Eco-Age, an integrated strategy consultancy, to evaluate their progress compared to others in the fashion industry. By using the scorecard, the company can see the areas that need improvement and also look to the future in setting new goals. Each year, Eco-Age assesses the retail business among a variety of different topics, ensuring industries take accountability and action when it comes to their performance. For each topic, Eco-Age rates the companies' activities where they can receive the following scores:

- No information = 0%
- Basics in Place = 1–24%
- Best Practice = 25–49%
- Leadership = 50–74%
- Setting New Standards = 75–100%

In 2021, Reformation received the highest scores on Sustainability Reporting & Transparency, Climate Action, and Low Impact Garment Care. On the opposite side of the spectrum, the company received the lowest scores on Biodiversity & Resource Efficiency, achieving Best Practice. Overall, the company continues to set high and impressive standards for the fashion industry (Reformation Sustainability Report 2021). Being transparent about your progress, even if it is not that flattering, lends to higher levels of trust in the public statements that a company makes.

Goals

The company's sustainability objectives fall within four pillars:

- made smarter (materials),
- made better (practices),
- made for good (circularity), and
- transparency.

According to the company website, the industry is responsible for creating **40 billion garments each year that are never sold**, and textile waste is projected to increase 60% by 2030. Reformation is working toward accomplishing their goal of sourcing 100% of their fabrics from recycled, regenerative or renewable materials made by 2025 to help mitigate their impact. The company wants to **use as close to 0% virgin materials**, only working with materials that are best for people and Earth. Another goal is to move towards a closed-loop system for all man-made cellulosic fibers to ensure emission control. The company will also look to work with suppliers to have 100% of materials receiving clean chemistry certifications, this will help to eliminate use and discharge of hazardous materials (Sustainability at Reformation 2023).

Embracing Circularity

Reformation believes that **circularity is the "only way forward for fashion."** Unlike companies partaking in the fast fashion industry, the company focuses on getting the best quality materials in a sustainable way, creating fashion that is meant to last longer. For example, Reformation worked with the Austrian company, Lenzing, to create regenerated cellulose fiber identical to cotton called TENCEL. The fiber is produced mainly from Eucalyptus trees, just needing about half an acre of land to create a ton of the fiber! Through its acceptable practices, TENCEL had a water footprint of 65% smaller than cotton. The closed-loop production process saves over 99% of non-toxic solvent and is reused in the system rather than being flushed out as wastewater (Better Materials 2023). Reformation also encourages its customers to recycle or resell their undesired clothing. Through their reuse program, RefRecycling, consumers have the option of getting paid while keeping their clothing in cycle forever. With the help of Supercircle, a waste management solutions company, Reformation breaks down the worn clothing into workable material and merges them into new products. Another route **customers can take is reselling their clothes through ThredUp**, an online thrift shop, and receive Reformation store credit. Through their programs, the company recycled 36,984 pounds of manufacturing waste saving 1.66 MT of carbon dioxide emissions in 2022 and, additionally, resold or recycled over 40,000 garments with ThredUp (The Sustainability Report Reformation 2022).

GREEN CHEMISTRY

It is found that 75% of sustainability impact within the fashion industry happens at the raw material phase. To address their green chemistry initiatives, Reformation launched **Ref Fiber Standard** considering a wide range of concerns such as "**water input, energy input, land use, eco-toxicity, greenhouse gas emissions, human toxicity, availability, and price**." The ratings are as follows:

- *A–All-stars*: Natural fibers that are rapidly renewable, plant-based and have a potential for circularity.
- *B–Better than most*: Fibers are almost all natural or recycled fibers.
- *C–Could be better*: Fibers are better alternatives than more commonly used fibers, but not as innovative.
- *D–Don't use unless certified*: Require certifications for raw material cultivation (i.e., organic), animal welfare, traceability or wet processing.
- *E–Eww, never*: Fibers are too environmentally or socially intensive, and don't meet the sustainability criteria.

(Better Materials 2023)

In 2022, 93% of fibers used received A/B ratings helping Reformation's clothing line look better, feel better, and live forever. Another issue that is of concern for apparel is dyeing fabrics, which requires the use of and generation of hazardous chemical waste and wastewater. To combat such detrimental externalities, Reformation has 84% of dyers and printers with clean chemical certifications, almost achieving their goal of 100%. Among the Tier 1 and 2 suppliers, they have achieved 100% traceability, knowing exactly who makes their clothes, not an easy undertaking (The Sustainability Report 2022).

REFORMATION GREENER PRODUCT PROGRAM

- Commitment to circularity by establishing collection and recycling of used clothing.
- Have a Sustainability Scorecard addressing sourcing, manufacturing, and their products.
- Use green chemistry and its standards to improve production processes.
- Set aggressive goals to drive higher sustainability performance of products, e.g., use as close to 0% virgin materials.

The fashion apparel industry has lots of room for improvement and is in need of changing their practices to become more sustainable. Reformation has committed to leading the industry toward a brighter and healthier future. There are a number of goals and policies that have been put in place to ensure that the future of fashion becomes more environmentally friendly with a big focus on circularity.

Common Practices among Leaders

By evaluating leading companies' activities, we can determine what the key initiatives used are to bring greener products to market. Based on the case studies presented in this chapter we see three main initiatives used:

- A robust framework for product developers.
- Setting goals for developing greener products.
- Communication process for product improvements.

FRAMEWORK FOR PRODUCT DEVELOPERS

Leading companies believe that it is imperative to have methods that make it as straightforward as possible for product developers to take strides to make their products greener. Just about every company evaluated has a program to enable greener product design. The most frequent focus areas for these programs include reduction of energy, removal of toxic materials, reducing the product size and weight, packaging, and use of life cycle thinking to identify the key aspects of a product to focus on environmental improvements. Also, the use of a scorecard is a leading practice that helps design teams to make the most meaningful improvements. Companies like Method, Philips, Seventh Generation, GE, and J&J all use focal areas and scorecards to help design teams to advance greener product design. In developing these frameworks, several companies have partnered with third parties to bring validity to their approach. Some have used consulting firms to assist in developing scorecards and others audits by third parties of their greener product approach. One of the keys to these scorecards is to keep it simple. If you initiate a complicated program that slows developer down, your greener product development program is not going to take off.

GOALS FOR DEVELOPING GREENER PRODUCTS

Just having a framework alone does not appear to be enough to bring greener products to market. Most companies use sustainability goals focused on product improvements. Examples are revenue goals for greener products and R&D spending on sustainable innovations which GE, P&G, and J&J have all set. Several companies have goals that focus on sustainable sourcing, improvement in energy efficiency, removal of problematic materials like PVC and other hazardous chemicals, and management of end-of-product life. The old adage, you manage what you measure, applies to greener product initiatives too. I have seen the use of score cards to be very effective in highlighting a product on track or not. Use of the stop light, green, yellow, red connected to metrics and data is a common method. It is a quick way to help management see which products are meeting the companies' objectives and which need encouragement to be better.

Communication Process

In order to inform customers of the positive attributes of their sustainable products, leading companies have developed ways to make it clear to customers that their products have improved performance. Branded green product lines like Ecomagination, Green Works, or Earthwards® enable customers to quickly identify a greener product. Some companies have used innovative methods to communicate improvements, others have looked outside their companies for assistance in communicating their greener products with third-party certifications such as the US EPA's Energy Star and Safer Choice certifications or the C2C and FSC sustainable forest certification. Another communication method to gain the trust of customers is the transparency of materials and ingredients used. Several companies have made bold commitments to transparency, agreeing to disclose every ingredient used. This is especially true for the deep green-based products like Method and Seventh Generation, but also with household product companies like SC Johnson and Clorox. Other innovative communication tools are product environmental reports that indicate various environmental and sustainability information. We will discuss these communication methods in more detail when evaluating the marketing programs companies use later in this book.

REFERENCES

Allianz Arena General Information. 2023. Allianz Arena. https://allianz-arena.com/en/arena/facts/general-information#:~:text=The%20new%20Philips%20lighting%20is,lighting%20sy. stem%20is%20very%20robust (Accessed August 12, 2023).

Apple Inc. 2010. A Greener Apple. http://www.apple.com/hotnews/agreenerapple/ (Accessed December 11, 2010).

Apple Inc. 2022. Apple Watch Ultra Environmental Report 2022. https://www.apple.com/co/environment/pdf/products/watch/Apple_Watch_Ultra_PER_Sept2022.pdf. (Accessed April 21, 2024).

Balch, O. 2023. Ryan Gellert: At Patagonia we have to be greener than green – and make a profit. Reuters. https://www.reuters.com/business/sustainable-business/ryan-gellert-patagonia-we-have-be-greener-than-green-make-profit-2023-04-11/ (Accessed July 23, 2023).

Base of the Pyramid Impact Report. 2021. SC Johnson. https://www.scjohnson.com/en/stories/healthier-world/mosquito/bop-impact-report (Accessed August 13, 2023).

BASF. BASF confirms ambitious climate targets and takes steps to reduce product-related emissions. 2022. https://www.basf.com/global/en/media/news-releases/2022/03/p-22-191.html (Accessed August 13, 2023).

BASF. Eco-Efficiency Analysis. 2023. BASF. https://www.basf.com/us/en/who-we-are/sustainability/we-drive-sustainable-solutions/quantifying-sustainability/eco-efficiency-analysis.html (Accessed August 13, 2023).

BASF Corporation. 2023. About BASF. https://www.basf.com/us/en/who-we-are.html (Accessed August 13, 2023).

BASF. Global Product Strategy (GPS). 2021. Vhttps://www.basf.com/global/en/who-we-are/sustainability/we-produce-safely-and-efficiently/safety/product-stewardship-and-global-product-strategy.html (Accessed April 21, 2024).

BASF. Risk Management Process. 2023. BASF. https://report.basf.com/2022/en/managements-report/forecast/opportunities-and-risks/risk-management-process.html (Accessed August 14, 2023).

BASF. Triple S. https://www.basf.com/global/en/who-we-are/sustainability/we-drive-sustainable-solutions/sustainable-solution-steering.html (Accessed April 21, 2024).

Better Materials. 2023. Reformation. https://www.thereformation.com/sustainability/our-stuff.html (Accessed August 13, 2023).

Beeson, L. 2021. Scientists find eco-friendly way to dye blue jeans. University of Georgia Today. https://news.uga.edu/scientists-find-eco-friendly-way-to-dye-blue-jeans/(Accessed August 10, 2023).

Biobased, Plant-based - What Do They Mean? 2018. Seventh Generation. https://www.seventhgeneration.com/blog/biobased-plant-based-what-do-they-mean (Accessed August 13, 2023).

Citizenship & Sustainability Report. 2015. Johnson & Johnson. p. 9. https://www.google.com/search?q=johnson+%26+johnson+sustainability+report+2015&oq=johns on+%26+Johnson+sustainability+rep&aqs=chrome.0.0j69i57j0l4.27326j0j8&sourceid=chrome&ie=UTF-8 (Accessed February 9, 2017).

Chouinard Y. 2023. Earth is now our only shareholder. Patagonia. https://www.patagonia.com/ownership/ (Accessed July 24, 2023).

Corporate Conscious Report. 2021. Seventh Generation. https://www.seventhgeneration.com/sites/default/files/2022-09/SVG_Fingerprints_Climate_Impact_Report_2021.pdf (Accessed August 13, 2023).

Corporate Conscious Update. 2020. Seventh Generation. https://www.seventhgeneration.com/company/goals-impact (Accessed August 13, 2023).

Demystifying EcoDesign – and the Impact When We Get It Right. 2020. Philips. https://www.philips.com/a-w/about/news/archive/blogs/innovation-matters/2020/20200729-demystifying-ecodesign-and-the-impact-when-we-get-it-right.html (Accessed July 29, 2020).

Disinfecting Multi-Surface Cleaner – Lemongrass Citrus. 2023. Seventh Generation. https://www.seventhgeneration.com/disinfecting-multi-surface-cleaner#:~:text=Other%20 ingredients%3A%20Aqua%20(water),cymbopogon%20nardus%20(citronella)%20 oil%2C (Accessed August 13, 2023).

EasyDose™ Ultra Concentrated Laundry Detergent – Free & Clear. 2023. Seventh Generation. https://www.seventhgeneration.com/easydose-ultra-concentrated-laundry-detergent-freeclear (Accessed August 13, 2023).

Ecomagination Ten Years Later: Proving that Efficiency and Economics Go Hand-in-Hand. 2015. General Electric Company. https://www.ge.com/news/reports/ecomagination-ten-years-later-proving-efficiency-economics-go-hand-hand (Accessed July 23, 2023).

Ecovio® (PBAT, PLA) – Certified Compostable Polymer With Bio-based Content. 2023. BASF. https://plastics-rubber.basf.com/global/en/performance_polymers/products/ecovio.html (Accessed August 13, 2023).

Environmental Progress Report. 2023. Apple. https://www.apple.com/environment/pdf/Apple_Environmental_Progress_Report_2023.pdf (Accessed August 23, 2023).

GE Digital Industrial Annual Report. 2015. General Electric Company. https://assets.ey.com/content/dam/ey-sites/ey-com/en_gl/topics/trust/GE_AR15_Integrated_Summary_Report.pdf (Accessed January 28, 2017).

GE Environment. https://www.ge.com/about-us/history#/narrative/environment (Accessed December 17, 2023).

GE Healthcare Names New 3.0T MRI System for Today's Healthcare Heroes. 2021. General Electric Company. https://www.gehealthcare.com/about/newsroom/press-releases/ge-healthcare-names-new-30t-mri-system-for-todays-healthcare-heroes?npclid=botnpclid(Accessed April 21, 2024).

Get Your Software Kicks on Predix: GE Opens the World's First Industrial App Marketplace. 2016. General Electric Company. https://www.ge.com/news/press-releases/ge-predix-software-platform-offers-20-potential-increase-performance-across-customer (Accessed April 21, 2024).

GE Sustainability Report. 2021. General Electric Company. https://www.ge.com/sites/
default/files/ge2021_sustainability_report.pdf (Accessed July 23, 2023).

Global Fashion Industry Statistics. 2023. FashionUnited. https://fashionunited.com/global-
fashion-industry-statistics (Accessed August 14, 2023).

Greenpeace. 2023. About Greenpeace. https://www.greenpeace.org/usa/about/#:~:text=
Greenpeace%20is%20a%20global%20network,a%20green%20and%20peaceful%20
future.&text=We.,Are (Accessed August 13, 2023).

9HA Gas Turbine. 2023. General Electric Company. https://www.ge.com/gas-power/
products/gas-turbines/9ha (Accessed August 13, 2023).

Health for Humanity 2025 Goals. 2022. Johnson & Johnson. https://healthforhumanityreport.
jnj.com/2022/our-approach/health-for-humanity-2025-goals.html (Accessed August
13, 2023).

How Starbucks Saves Millions a Year in Energy with LED Lighting. 2016. GreenBiz. https://
www.greenbiz.com/blog/2010/12/02/how-starbucks-saves-millions-year-energy-led-
lighting (Accessed September 27 2016).

Iannuzzi, Al. 2012. *Greener Products: The Making and Marketing of Sustainable Brands.*
CRC Press, New York, p. 50, 51, 59, 98.

Iannuzzi, Al. 2017. *Greener Products: The Making and Marketing of Sustainable Brands.*
CRC Press, New York, p. 80.

Integrated Annual Report. 2016. The Clorox Company. https://investors.thecloroxcompany.
com/investors/news-and-events/press-releases/press-release-details/2016/Evolve-This-
Clorox-2016-Integrated-Annual-Report-Highlights-Fiscal-Year-Progress-Against-
2020-Strategy-and-More-Than-a-Century-of-Evolution-in-Health-and-Wellness/
default.aspx(Accessed April 21 2024).

Integrated IGNITE ESG Goals and Priorities. 2023. The Clorox Company. https://www.
thecloroxcompany.com/company/ignite-strategy/about-ignite-esg/ (Accessed August
13, 2023).

Jeong, J.E. 2023. Samsung Electronics gets the highest-level certification from AWS. The
Korea Economic Daily. https://news.samsungsemiconductor.com/global/samsung-
semiconductor-site-awarded-highest-level-global-certification-for-water-resource-
management/ (Accessed April 21, 2024).

Johnson & Johnson Earthwards®: The Unique Johnson & Johnson Program That's Helping
to Create a More Sustainable World. https://www.jnj.com/innovation/earthwards-
a-johnson-and-johnson-program-helping-create-a-more-sustainable-world (Accessed
May 12, 2024).

Journey to a More Sustainable, Equitable Planet: Seventh Generation's 2025 Goals.
2023. Seventh Generation. https://www.seventhgeneration.com/blog/journey-more-
sustainable-equitable-planet-seventh-generations-2025-goals (Accessed August 13,
2023).

Materials. 2023. Patagonia. https://www.patagonia.com/materials/ (Accessed July 23,
2023).

Mazur, N.F. 2017. Earthwards®: The Unique Johnson & Johnson Program That's Helping
to Create a More Sustainable World. Johnson & Johnson. https://www.jnj.com/
innovation/earthwards-a-johnson-and-johnson-program-helping-create-a-more-
sustainable-world (Accessed August 13, 2023).

Method Products Packaging. 2023. Method Products. https://methodproducts.co.uk/beyond-
the-bottle/packaging/ (Accessed August 13, 2023).

Mohan, A.M. 2022. P&G uses Life Cycle Assessment for Sustainable Packaging
Design. Packaging World. https://www.packworld.com/news/sustainability/article/
22340271/pg-uses-life-cycle-assessment-for-package-design (Accessed August 13,
2023).

MR Ingenia Ambition 1.5T. 2023. Philips. https://www.usa.philips.com/healthcare/resources/landing/the-next-mr-wave/ingenia-ambition?utm_id=71700000097929390&origin=7_700000002330341_71700000097929390_58700008240171761_43700074871892778&gclid=CjwKCAjw_uGmBhBREiwAeOfsdxtUyYr3oXKE7wee6-oe-aPsrsm1fnHJ5WpoyFOUaaE3NbJfCVWjHBoCyQ0QAvD_BwE&gclsrc=aw.ds (Accessed August 13, 2023).

Organic Cotton. 2023. Patagonia. https://eu.patagonia.com/hr/en/our-footprint/organic-cotton.html (Accessed August 13, 2023).

Our Environmental Responsibility Program. 2023. Patagonia. https://www.patagonia.com/our-responsibility-programs.html (Accessed July 23, 2023).

Our Purpose & ESG Commitments. 2023. Philips. https://www.philips.com/a-w/about/environmental-social-governance/our-purpose.html (Accessed August 13, 2023).

Philips Achieved All Sustainability Targets. 2021. Philips. https://www.philips.com/a-w/about/news/archive/standard/news/press/2021/20210223-philips-meets-its-healthy-people-sustainable-planet-targets-and-forges-ahead-with-integrated-esg-framework.html (Accessed August 13, 2023).

Planet & Society. 2023. Unilever. https://www.unilever.com/planet-and-society/#:~:text=We're%20a%20company%20of,to%20make%20sustainable%20living%20commonplace (Accessed August 13, 2023).

Preda, M. 2023. Trust the Scientists. Patagonia. https://www.patagonia.com/stories/trust-the-scientists/story-94032.html (Accessed July 23, 2023).

Procter & Gamble. Citizenship Report 2022 – Environmental Sustainability. 2022. https://us.pg.com/citizenship-report-2022/environmental-sustainability/ (Accessed August 13, 2023).

Product Environmental Report iPhone 14. 2022. Apple. https://www.apple.com/environment/pdf/products/iphone/iPhone_14_Pro_PER_Sept2022.pdf (Accessed August 13, 2023).

Products & Packaging Brand Highlights. 2023. The Clorox Company. https://www.thecloroxcompany.com/responsibility/clean-world/products-packaging/ (Accessed August 13, 2023).

Product Stewardship/Earthwards. 2023. Johnson & Johnson. https://www.jnj.com/global-environmental-health/product-stewardship-earthwards (Accessed July 28, 2023).

Public Sustainability Report. 2015. SC Johnson. https://corp-ucl.azureedge.net/-/media/sc-johnson/our-purpose/sustainability-reports/2015/sc-johnson-2015-public-sustainability-report-en.pdf (Accessed October 15, 2016).

Reducing Emissions from the Use of Our Products. 2023. Unilever. https://www.unilever.com/planet-and-society/climate-action/reducing-emissions-from-the-use-of-our-products/#:~:text=We're%20using%20our%20influence,and%20by%205%25%20since%202021 (Accessed August 13, 2023).

Refill + Reuse. 2023. Method Products. https://methodproducts.com/method/meet-concentrates/ (Accessed August 13, 2023).

Reformation. 2023. About-Us Reformation. https://www.thereformation.com/sustainability/oh-hi.html (Accessed August 13, 2023).

Responsible Recycling. 2023. Samsung. https://www.samsung.com/us/explore/sustainability/responsible-recycling/ (Accessed August 13, 2023).

Responsibility Standards for Suppliers. 2023. Johnson & Johnson. https://www.jnj.com/suppliers/responsibility-standards-for-suppliers (Accessed April 21, 2024).

Rethinking Plastic Packaging. 2023. Unilever. https://www.unilever.com/planet-and-society/waste-free-world/rethinking-plastic-packaging/ (Accessed August 13, 2023).

Rockeman, O. 2023. Reformation Eyes Going Public in Testing the Limits of Eco-Friendly Fashion. Bloomberg. https://www.bloomberg.com/news/features/2023-05-09/sustainable-fashion-brand-reformation-eyes-ipo-as-sales-top-300-million?in_source=embedded-checkout-banner (Accessed August 15, 2023).

Safe Returns. 2023. Simponi Golimumab. https://www.simponi.com/rheumatoid-arthritis/safe-returns (Accessed August 13, 2023).

Samsung Electronics Announces 2022 Initiatives to Make Home Appliances More Eco-Conscious. 2022. Samsung. https://news.samsung.com/global/samsung-electronics-announces-2022-initiatives-to-make-home-appliances-more-eco-conscious (Accessed August 13, 2023).

Samsung Electronics Sustainability Report. 2023. Samsung Electronics. https://www.samsung.com/global/sustainability/media/pdf/Samsung_Electronics_Sustainability_Report_2023_ENG.pdf (Accessed August 13, 2023).

Samsung Repurposes Discarded Fishing Nets For New Galaxy Devices. 2022. Samsung. https://news.samsung.com/us/samsung-repurposes-discarded-fishing-nets-new-galaxy-devices-unpacked-2022/ (Accessed August 13, 2023).

SC Johnson Greenlist Program. 2017. SC Johnson. https://www.scjohnson.com/Our%20Purpose/Sustainability%20Report/Explaining%20the%20SC%20Johnson%20Greenlist%20Program%20An%20Excerpt%20from%20Our%202017%20Sustainability%20Report (Accessed August 13, 2023).

Seventh Generation: More Than a Name. 2022. Seventh Generation. https://www.seventhgeneration.com/blog/more-than-a-name (Accessed August 13, 2023).

Staying ahead of PFAS regulations in the industrial coatings market. 2023. BASF Insights. https://insights.basf.com/home/article/read/staying-ahead-of-pfas-regulations-in-the-industrial-coatings-market (Accessed September 6, 2023).

Supplier Sustainability Performance. 2023. Philips. https://www.philips.com/a-w/about/environmental-social-governance/environmental/supplier-sustainability/supplier-sustainability-performance.html (Accessed August 13, 2023).

Sustainability at Reformation. 2023. Reformation. https://media.thereformation.com/image/upload/v1645172843/pdfs/Reformation-Sustainability-Framework_2022.pdf (Accessed August 14, 2023).

Sustainability Report. 2017. SC Johnson. https://corp-uc1.azureedge.net/-/media/sc-johnson/our-purpose/sustainability-reports/2017/scjohnson2017sustainabilityreport.pdf (Accessed July 24, 2023).

Sustainable & Deforestation-Free Palm Oil. 2023. https://www.unilever.com/planet-and-society/protect-and-regenerate-nature/sustainable-palm-oil/ (Accessed August 13, 2023).

Sustainable & Regenerative Sourcing. 2023. Unilever. https://www.unilever.com/planet-and-society/protect-and-regenerate-nature/sustainable-and-regenerative-sourcing/ (Accessed August 13, 2023).

The Race to Zero Waste: Seventh Generation's Journey. 2019. Seventh Generation. https://www.seventhgeneration.com/blog/race-zero-waste-seventh-generations-journey (Accessed August 13, 2023).

The Sustainability Report: 2021 Year in Review. 2021. Reformation. https://www.thereformation.com/sustainability/oh-hi.html (Accessed August 14, 2023).

The Sustainability Report 2022. 2022. Reformation. https://media.thereformation.com/image/upload/v1683921070/pdfs/Reformation-Annual-Report-2022.pdf (Accessed August 13, 2023).

The Unilever Compass. 2023. Unilever. https://www.unilever.com/files/8f9a3825-2101-411f-9a31-7e6f176393a4/the-unilever-compass.pdf (Accessed August 13, 2023).

The World's Food System Needs to Change. 2023. Unilever. https://www.unilever.com/brands/nutrition/the-worlds-food-system-needs-to-change/ (Accessed August 13, 2023).

Thriving Together: Burt's Bees 2020 Impact Report. 2020. The Clorox Company. https://www.thecloroxcompany.com/blog/thriving-together-burts-bees-2020-impact-report/ (Accessed August 13, 2023).

Ultramid® Biomass Balance (BMB). 2023. BASF. https://chemicals.basf.com/global/en/Monomers/polyamide_intermediates/sustainable_polyamide_solutions/ultramid_biomass_balance.html (Accessed August 14, 2023).

Water Stewardship. 2023. Unilever. https://www.unilever.com/planet-and-society/protect-and-regenerate-nature/water-stewardship/ (Accessed August 13, 2023).

Worn & Wear. 2023. Patagonia. https://wornwear.patagonia.com/ (Accessed July 24, 2023).

Zero Plastic Line. 2021. Seventh Generation. https://www.seventhgeneration.com/blog/zero-plastic-line#:~:text=It's%20our%20revolutionary%20new%20line,or%20plastic%20of%20any%20kind (Accessed August 13, 2023).

5 Ways Companies Can Weave Sustainability Into Their DNA. 2016. General Electric Company. https://www.ge.com/news/taxonomy/term/2747 (Accessed August 13, 2023).

5 Sustainable Innovation and Tools for Product Sustainability

James Fava

There has been an explosion of interest, commitment, and actions to address sustainability issues across the board. Over the last five years, this explosion has included solid interest in greener products, driven by new and expanded government, business, and society's expectations. The previous chapters in this third edition have done an excellent job of summarizing and explaining these. While there are many paths that could be taken within this chapter, the option that provides an amazing opportunity is the interest in embedding sustainability attributes into innovation. The next frontier is the embedment of sustainability (including life cycle) information into innovation as the scale needed to enable the innovation to occur at the level necessary to address the global issues facing us (e.g., climate change, circular, biodiversity, social welfare).

What do we mean by Innovation? To me, I have seen innovation change greatly over my 45+ years working in this field. My dissertation was typed on an IBM Select typewriter. We could go back 1 space to white-out a mistake. Now we have Artificial Intelligence (AI), social media allows us to consume real-time information including events around the world, and products are being designed more and more with sustainability attributes embedded into their design. All that because of innovation. The next frontier is the collaborative effort to further embed sustainability into innovation. Thus, we titled this chapter to be Sustainable Innovation and Tools for Product Sustainability.

This chapter will highlight three meaningful and impactful examples and insights: case studies on the application of product sustainability tools, Golden Rules for Sustainable Business, and a call for action to create more sustainable products.

A FRAMEWORK FOR GREENER PRODUCTS EXISTS

Emerging issues and market expectations that companies face are increasing and expanding. The strategy implementation framework for greener products allows for clear, conscious, deliberate definition of an environmental strategy with an understanding of the tools and systems that the strategy implies and necessitates, providing a framework within which emerging issues can be managed. Its foundations are a clear strategic intent and have been segmented into four-strategy options: *compliant*, *market-driven*, *engaged*, and *shaping the future*. These strategy options evolved into Golden Rule 7 – *Meet them where they are*, which will be expanded on later in this chapter.

DOI: 10.1201/9781003441939-7

Maturity models create a framework to define "how" to improve products. With any emerging or existing sustainability issue, several questions need to be asked: (1) What is the issue? (2) Is it relevant to my company, regions, and/or products – the why? (3) How do I respond? The ultimate success factor is linking product sustainability and business value. Product Innovation Tools such as the Reckitt Benckiser's Sustainable Innovation App, which is known as Footprinter's Green Design tool, is reviewed. Application of Life Cycle Impact Hot Spots Analysis to Inform Greener Products is discussed.

Collaboration efforts within the industry can advance sustainability. Groups that demonstrate the practices and outcomes of an effective collaboration platform are the Electric Utility Industry Sustainable Supply Chain Alliance (SSCA), the Alliance and the Product Sustainability Round Table (PSRT), the Sustainable Apparel Coalition, and the Sustainable Purchasing Leadership Coalition. By participating in a collaboration platform, individuals and organizations are able to scale resources around noncompetitive efforts, share successful practice, and build on mutual lessons learned.

Managing product stewardship can be done with in house resources or managed services. As companies look to downsize or "right size" as it has sometimes been called, a question that often surfaces is, how will we continue to meet the demands of the legal requirements and market demands. At the end, it is all about change management – embedding a product sustainability tool – a change management success example is the Earthwards® Framework for Developing Greener Products (Fava 2018) sets the stage that tools, standards, and frameworks exist to embed sustainability considerations into the design, development, and commercialization of products. As has been described earlier in this book, the emerging issues and market expectations that companies face are increasing and expanding. The strategy implementation framework (Figure 5.1) initially presented in Fava (2012) allows for a clear, conscious, deliberate definition of an environmental strategy with an understanding

FIGURE 5.1 Strategy implementation business framework. (From Fava, Framework for developing greener products, in Al Iannuzzi, *Greener Products: The Making and Marketing of Sustainable Brands*, 1st edn, CRC Press, Boca Raton, FL, 2012. With permission.)

of the tools and systems that the strategy implies and necessitates providing a framework within which emerging issues can be managed.

These framework and strategy levels provide a flexible framework of systems, programs, and tools. Because the goals of various tools overlap, it is possible to link these approaches on a situation-specific basis through a flexible conceptual framework.

The evolution of environmental management strategies within businesses has been influenced by external drivers (e.g., regulations and customer requirements), internal drivers (e.g., cost savings and environmental strategies), and internal capacity and resources. As firms move from a compliant strategy to a more sustainable strategy, different implications result (see Golden Rule 7 for more information). A *compliant* **strategy**, for example, is often viewed as a cost and often includes only strategic elements aimed at meeting the legal requirements as efficiently as possible. In a *market-driven* **strategy**, for example, a firm has integrated pollution prevention and customer/consumer or reactive market considerations into the design of its products or processes, which results in cost savings and/or cost avoidance. On the other hand, *shaping the future* **strategy** may generate revenue by viewing the environment from a strategic perspective to identify new business opportunities and greener products. One advancement has been building alignment of the strategy levels and implementation of frameworks to various business values, that is, growing revenue, enhancing brands, reducing costs, and mitigating risks. This is further elaborated in the Golden Rule 7 – *meet them where they are* section, later in the chapter.

This chapter builds on and demonstrates **applications of the framework** toward a better understanding of how companies have begun to learn from these key insights and apply the tools and approaches into their core business practices. There have been several advancements to the product sustainability framework as companies have made significant process implementation:

- *Product innovation tools*: As has been discussed throughout this book and *Greener Products* (Iannuzzi 2012), there has been an ongoing advancement in the development and application of product innovation tools. Two examples are highlighted. First, **Reckitt Benckiser (RB) has developed a "Green Design tool"** they call **"Sustainable Innovation App"** which can be considered a hybrid since it combines traditional quantitative life-cycle assessment (LCA) methods with qualitative "**score carding**." Often, a company has within their toolbox tools to ensure compliance with government regulations. Additionally, where there is market demand for broader sustainability information (e.g., energy, greenhouse gases, water) from companies, LCA is the right tool (e.g., GaBi or SimaPro, Footprinter). The RB Green Design tool combined the product compliance issue with this broader life-cycle information into one tool. **Second**, the appliance sector developed and applied a "**hot-spots analysis (HSA)**" framework as a foundational step toward the development of a series of product sustainability standards that are being used to inform innovation teams on criteria which should be met for a product to be considered "greener."

- *Maturity models create framework to define "how" to improve*: With any emerging or existing sustainability issue, a number of questions need to be asked:
  1. *What* is the issue?
  2. Is it relevant to my company, regions, and/or products? (The *why*)
  3. *How* do I respond?

     As can be seen through the review of social media, many companies have begun to understand the "what and the why" (although there is still much work to be done here). The challenge is "how do I respond." Each company may be at a different place along a continuum from ad hoc or just starting a sustainability journey to those that have already culturally embedded sustainability criteria into their business practices. These different levels of maturity are being recognized, and maturity models are being developed to inform a company of the various elements that should be included within their own strategy implementation framework and different levels of maturity.

- *Collaboration to advance performance and build capacity*: As a company moves along maturity levels, there can be topics or situations where working with other companies can influence the performance of each company but also the sector. The Electric Utility Industry Sustainability Supply-Chain Alliance (Alliance) is one of these collaborations that has made excellent progress toward making their supply chain more sustainable.

- *Engaging the right resources to do the job: Managed services*: Any framework and toolbox are only as good as the people or resources that use or apply them. Companies are continually being asked to optimize resources against revenue and other business metrics. Application of managed services or outsourced services to certain (but not all) sustainability practices has surfaced.

- *At the end it is all about change management–embedding a product sustainability tool–change management for success*: Many tools and data systems have been developed and remain unused or underutilized. Often problems develop because adjusting or changing an existing business practice would be required. The case study using Johnson & Johnson's successful implementation of its Earthwards® approach provides interesting insights and learning on change management success factors.

- *The ultimate success factor–linking product sustainability and business value*: Since the publication of the earlier version of this chapter, five years ago in *Greener Products* (Iannuzzi 2018), one of the most interesting and compelling advancements is an update of the simple graphic in Figure 5.2 – business implications on the strategy implemented. The ability to translate any sustainability plan, targets, and/or goals into the language of the decision-makers has been accelerated through the use of the business value framework. **Does your greener product tool/plan contribute to growing revenues, enhancing your brand, reducing costs, or mitigating risks?**

For each of these advancements to the original sustainability implementation framework, the following examples and insights are provided to inform the reader as to best practices and steps they can take to culturally embed sustainability.

PRODUCT INNOVATION TOOLS

Streamlined Product Innovation tool—The previous chapter described the various LCA approaches that exist today. Early adopters had focused on applying LCA within the development process, for example, Daimler and BMW. They provided excellent examples of how life-cycle information can be used to inform design. Within the product stewardship space, there is a continuum for ensuring that products are in compliance with governmental legal *requirements* to ensure a company's ability to sell their products within the countries they operate. Additionally, there *is a* growing demand for information on the environmental and social impact of products over their entire life cycle. Companies are looking for tools that address the continuum from product compliance to broader life-cycle information. RB has developed a "hybrid" tool that does just that.

RECKITT BENCKISER'S SUSTAINABLE INNOVATION APP DESCRIPTION OF THE PROBLEM

RB, one of the global health, hygiene, and home leaders, is all about performance. The company's betteRBusiness strategy has some big 2020 goals, including a third of their net revenue coming from more sustainable products and a third reduction in carbon and water impact per dose. To achieve this, RB recognized that everyone in the business had a role to play. They needed to tie these corporate-level targets back to product design.

RB's definition of "**more sustainable**" includes quantitative (LCA-like) assessment of carbon and water impacts and qualitative "attributes" such as an FSC-certified paper board. RB wanted to develop a tool that would be simple, easy, and accessible to everyone. A tool that combines qualitative and quantitative assessment gives real-time easy-to-understand results and delivers data/results in a manageable way that enables reporting and actionable insight.

RB began its journey in 2009 with a simple spreadsheet-based calculator that allowed innovators to understand the carbon impact of their design decisions. This helped RB achieve its 2020 target to reduce the carbon footprint of products by 20% per dose eight years early. It also highlighted limitations – RB could not explore broader environmental impacts and evaluate trade-offs (e.g., carbon versus water), and it was heavy on administration. They ended up each quarter with several hundred spreadsheets in an email inbox and a huge task to pull it all together into reports, let alone generate insights for action.

The Solution

To scale RB's sustainable product development program to the demands of the business and incorporate additional metrics (e.g., water) that provide a broader view of the opportunities for innovation, RB worked with one of the authors and the software company, Footprinter, to develop a web-based solution to facilitate sustainable

FIGURE 5.2 Sustainable Innovation App Screenshot, Footprinter.

innovation. A screenshot from the RB App (as viewed internally) is illustrated in Figure 5.2.

The Sustainable Innovation App is a web tool, allowing access to the latest version, by all R&D staff, wherever they are. It has a user-friendly interface that allows users to carry out assessment quickly and easily. Users build up their product by searching for raw and packaging materials and filling out details such as functional unit, material weight, recycled content, and consumer-use activity. Tags are applied to each assessment, such as product category and brand. This allows for centralized roll-up and analysis across assessments. Users compare the impacts of their product against a "benchmark" product. Red/Amber/Green gauges show users how their product performs against the benchmark in real time. Overall, pass/no-pass simplifies results.

The tool is then embedded into RB's new product-design process, and designers have to assess innovations in order to progress through key stage gates. This is critical as it has meant that sustainable thinking and design has been integrated into "Business as Usual," and it is incorporated *early* in the design process. **All the RB R&D teams are trained to use the tool**; "champions" within each team are given extra training and carry out reviews/checks.

The engine that powers RB's Sustainable Innovation App is known as Footprinter's Green Design tool (Footprinter 2018). It is a framework which is customized for each company and can be launched in as little as two months. While RB is particularly interested in measuring the carbon and water impact during consumer-use (a significant hot spot for their products), the tool has also been configured for a major tech hardware manufacturer to focus on the EU circularity metric. RB (and others) get their (non-sustainability specialist) R&D teams to use the tool. When being configured for non-specialist use, **the tool is kept simple** with relatively little flexibility. Other companies employ a more complex version of Footprinter, this is used by internal sustainability teams as a customized LCA tool, and it has been verified to produce ISO 14040/14044 compliant LCAs.

The tool's flexibility means it has been used to implement some non-standard LCA metrics, such as "naturalness" (% of raw and packaging materials from natural sources) and the EU circularity metric. It also allows for combinations of quantitative and qualitative metrics, such as RB's "ingredients" metric where users qualify whether products have certain sustainability "credentials," such as FSC paperboard or Fairtrade certification.

Benefits – Sustainable and Business Value

RB's previous spreadsheet-based calculator was passed around its champion network to manually enter details of their innovations periodically for the sustainability team's review. This worked well, but it isolated insights with just a few of RB's team and wasn't always accessible early on in the design process.

RB's Sustainable Innovation App surpassed their previous solution in a number of ways:

- Enables designers to innovate and the sustainability team to engage and support, not just validate.
- Intuitive centralized online tool with searchable product library.
- Simple, visual, real-time traffic light indicators for each metric: carbon, water, packaging, and ingredients (sustainable attributes).
- Live comparison of results with other products/brand averages.
- Ability to quickly modify existing product assessments, avoiding entering new data each time.
- Hotspot-based focus for targeted innovations around the impacts that matter most.
- View and report on status of brands and overall portfolio and share results across the business.

Key Success Factors/Learnings

The Sustainable Innovation App was assured by the consulting firm EY and launched in November 2013. Within two months, more than half of RB's global research & development team were using it, with more than 700 product analyses completed. **The tool is now fully embedded in RB's New Product Development (NPD) process.** All new products have to be assessed through the app in order to progress through RB's NPD stage gates. At point of publication, the tool has 900 users with over 2,500 products analyzed. The results are used each year to report against RB's targets, and it is also third-party assured. RB is on track to meet the target of a third of their net revenue from more sustainable products.

The success of individual product assessments has led RB to expand the app's use into corporate level reporting. The tool is now used to complete LCAs for a basket of hundreds of "representative products" across RB's full portfolio. These representative products are then scaled up with sales data to estimate the total carbon and water impacts embodied in raw and packaging materials and from consumer use. This approach has been **third-party assured** and gives RB actionable insight into the sources of their impacts.

RB's goal was to have assessments take place in less than 30 minutes by non-LCA experts. RB has proven this is possible through the development of the Sustainable Innovation App.

APPLICATION OF HOT-SPOTS ANALYSIS (HSA) TO INFORM GREENER PRODUCTS

The **appliance sector** has worked together to develop a series of product sustainability standards to drive improved performance. A key element of this effort was the development and use of HSA to inform decision-making. Specifically, to identify those impacts beyond energy efficiency in the use stage based upon scientific information, results of LCA studies, professional and technical expertise, and stakeholder input.

Before we describe the appliance sector's experiences with developing and applying product sustainability standards and HSA, it would be useful to explain what is meant by HSA. UNEP (2014) described HSA as "a methodological framework that allows for the rapid assimilation and analysis of a range of information sources, including life cycle based and market information, scientific research, expert opinion and stakeholder concerns." The flagship project 3a of the UNEP/SETAC Life-Cycle Initiative titled "Hot spots analysis – global guiding principles and practices" has developed a methodological framework (UNEP 2017) to guide others on general considerations if one is interested in conducting an HSA. The methodological framework outlines eight steps to follow (Figure 5.3).

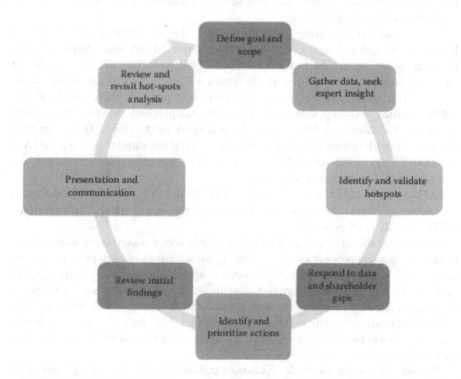

FIGURE 5.3 Methodological framework. (From UNEP, UNEP/SETAC life-cycle initiative – Flagship Project 3a (Phase 2), *Hotspots Analysis: An Overarching Methodological Framework and Guidance for Product and Sector Level Application*, Fava, J. and Barthel, M., co-chairs, May 2017; Paris, France. http://www.lifecycleinitiative.org/new-hotspots-analysis-methodological-framework-and-guidance/.)

The project team received feedback and insights from stakeholders (UNEP 2014):

- Support "Beyond LCA" approach (Quality + Quantity analysis).
- Clearly identify goal and scope.
- Build stakeholder credibility and use a phased approach.
- Keep hot-spots actionable and manageable.
- Prioritize to meet goals for addressing hot spots.
- Make results clear and intuitive using visualization.
- Periodically review and revise HSA.
- Clearly communicate uncertainty.
- Develop case studies and examples to support use.

The appliance sector efforts to use HSA to inform its development of product sustainability standards was one of over 40 examples of HSA the project team encouraged during its initial Phase 1 mapping of existing HSA approaches. The initiatives in the appliance sector were interesting as they established a set of sustainability standards for products. These standards serve as guidelines for companies, informing their practices in innovation.

As was realized through the results of the UNEP/SETAC Life-Cycle Initiatives Flagship Project (FS) 3a project report, **HSA goes beyond LCA**. In other words, it can be based on more information than an LCA study but can and often includes additional information from technical and professional experts. The appliance sector was able to utilize product life-cycle data, scientific studies, existing standards, industry and product experts, stakeholder concerns, and feedback from value chain players to develop a HSA that then informed the development of product sustainability standards (e.g., AHAM 7001-2012/CSA SPE-7001-12/UL. 7001, The Sustainability Standard for Household Refrigeration Appliances.

(AHAM 7001-2012/CSA SPE-7001-12/UL. 7001 2012). Three methodological advancements stand out as valuable to the work of the sector, but also influenced the UNEP/SETAC HSA project.

First, one of the advances used by the appliance sector was a collaborative stakeholder engagement process which combined characteristics of both private and consensus approaches. The resulting collaborative approach optimized the strengths of each approach to achieve a common goal to ensure stakeholder input and expert judgment, including testing of the draft standard prior to accreditation (Figure 5.4). The collaborative approach used provided a high level of credibility and stakeholder engagement, was highly transparent and relatively quick in its development of a standard and kept costs down.

Second, the overall life-cycle screening (including the HSA, scoping-level LCA, weighting) indicated that energy consumed during the use phase of the refrigeration appliance creates the largest environmental impact; as such, this unit process was identified as a high-priority hot spot. The life-cycle screening also identified four medium-priority hot spots related to the raw materials, manufacturing, and

Private	Consensus – based
Limited stakeholder input	*High-level of stakeholder input*
Low credibility	*High credibility*
Limited transparency	*Highly transparent*
Not typically accredited	Often accredited
Relatively short time to complete	Time intensive
Lower cost to develop	Higher cost to develop

Collaborative approach is a mix of two approaches

• Optimizes strengths of each approach to achieve common end goal
• Stakeholder input + expert judgment
• May involve testing of draft standard
  prior to accreditation

FIGURE 5.4 A collaborative approach.

end-of-life phases of the appliance. The corresponding attributes to encompass these hot spots are:

• Materials
• Energy during use
• Manufacturing and operations
• Product performance
• End-of-life management

The most significant deviation from the LCA results is that the Energy Consumption during use attribute was allotted 45% of the weighting within this Standard by the authors and the appliance standards team, compared to the greater than 70% share of life cycle impacts according to the LCA. The team arrived at this value because it represents the largest share of the attributes (consistent with the LCA), while taking stakeholder input into account and encouraging manufacturers to make improvements in the areas covered by the other attributes.

Third, HSA informed the prioritization of criteria that were used to develop specific targets which must be met in order for a new product to conform to the standard (e.g., AHAM 7001). But as was found to be relevant within the HSA methodological framework, the HSA informs on control or influence, that is, improve product performance in each criterion, thus resulting in a more sustainable product using the standard. This is exactly what happened, for example, Whirlpool was able to certify its KitchenAid 25 cu. Ft. 36" width refrigerator to the AHAM 7001-2012 standard (UL 2012).

This example is an illustration of an advancement of the product sustainability toolbox to include development in the HSA and its application within the appliance sector.

COLLABORATION TO ADVANCE PERFORMANCE AND BUILD CAPACITY

Sustainability by its very nature is complex and difficult to operationalize and implement. Multiple sectors and responsibilities are working to address this through collaboration platforms. A few examples include The Sustainable Apparel Coalition, Together for Sustainability, the Retail Industry Leadership Association, and the Sustainable Purchasing Leadership Coalition. By participating in a collaboration platform, individuals and organizations are able to scale resources around noncompetitive efforts, share successful practice, and build on mutual lessons learned. In practice, participation in effective collaboration platforms leads to:

- An improved understanding of key issues and opportunities through both shared experiences and resources.
- Accelerated performance from building on peer group models and lessons learned.
- Development of tools and resources to codify successful practice.
- Opportunities through benchmarking to identify risks and opportunities, as well as set priorities based on demonstrated value propositions.
- Opportunities to pursue more ambitious and impactful change by working with other like-minded organizations.

Two groups which demonstrate the practices and outcomes of an effective collaboration platform are the **Electric Utility Industry SSCA or the Alliance and the PSRT**. Both groups have an active membership group of approximately 15 member companies; they meet multiple times a year, operate within a **precompetitive space**, have an explicit culture of collaboration, and are acutely focused on driving impact reductions through multiple channels. Below, we unpack the success and lessons learned from the Alliance as a model for individuals and organizations to seek out, develop, and leverage to accelerate their sustainability journey.

The Alliance, over the last 3–4 years, has driven a tangible change within the industry, with members incorporating sustainability into decision-making, suppliers shifting from "check the box" levels of effort to innovative solutions, and a clear embracing of sustainability as a business priority throughout the supply chain. The Alliance's vision is "to be known as the leader in establishing a robust and sustainable electric utility industry supply chain, including advancing the maturity level of our members and stakeholders." It is supported by the supply-chain organizations of 15 leading electric utilities in the United States, including PG&E, DTE, Con Ed, Exelon, APS, Southern California Edison, etc. As a voluntary standards development organization, the development, implementation, and adoption of the standards are the anchor of all the Alliance's activities. The success of the Alliance has been driven by multiple factors including committed members, a dedicated executive director and administrative support, a supportive supplier community, etc. However, five elements really stand out as being critical success factors. They are, **a three-year strategic plan, effective standards, a sustainability framework, an executive-level dashboard, and an engaged community of practice**. Information on these elements can be found on the SSCA website thessca.org (SSCA 2023a).

The **three-year strategic plan** provides a common understanding of the direction, priorities, and goals of the group. This is complemented by yearly "roadmaps" that allow for adjustments to the pathway to the goals and are focused on execution. The **standards provide** a mechanism to document and codify successful practice on key material issues. A lot of effort is put toward ensuring the usability of the standards. The combination of these two factors creates effective standards that stakeholders want, and are able, to apply quickly within their organizations.

The **sustainability framework**—and associated benchmarking against it—helps to illuminate the internal processes and practices needed to be successful in this space. Further, by applying a maturity-model approach, the logical steps between "initiating" and "transformational practice" are clarified. Finally, by having a strategic goal linked to progress against the framework, the members are motivated to improve and the Alliance, through the community, is empowered to support those efforts.

The **executive-level dashboard** has proven to be a critical key in unlocking the engagement of senior leadership, as well as focusing member activities on areas of need. It has provided a mechanism for Chief Procurement Officers to quickly understand their performance relative to their peers on a variety of initiatives within the Alliance. Further, by being able to inform members of the scope of activities, they are similarly able to identify and focus on priority areas quickly (Figure 5.5).

Goal status	Goal description
	Member maturity improvement
	Releasing three voluntary standards
	New practice adoption
	100% survey users embed results into supplier relationships
	Continually release materials to educate stakeholders
	Speak at four events per year
	Increase traffic to our website

On target	Off track with recovery plan	Off track, target will be missed

FIGURE 5.5 Sample SSCA 2018 2018 goal progress dashboard. (From SSCA, state of the Alliance 2016, Presentation, *SSCA Supply Chain Sustainability Conference*, New Orleans, LA, September 29, 2016. With permission.)

Finally, and fundamentally, critical to the success of the Alliance is the **engaged community of practice**. With members willing to share both their challenges and successful practices around the standards, the community is able to advance their practices, adoption of the standards, and realization of results more efficiently.

The key lessons learned over this time have been to build momentum, focus, on small wins that you have control over, build on those with increasing ambition, understand and address the obstacles to improved performance, and engage leading members on direction setting while supporting those who are not as advanced in order to leverage the resources of the community.

The Alliance provides an effective model that other industries, organizations, and practices should consider. Regardless of where one is on the sustainability journey, participation in a collaboration platform similar to the Alliance or PSRT can be a powerful catalyst to speeding and scaling one's activities and performance.

MATURITY MODELS CREATE A FRAMEWORK TO DEFINE "HOW" TO IMPROVE

As emerging issues are identified and evaluated as to their relevance to a business, the business needs to understand **what** they are and **why** they are relevant or not. Once they understand the **what** and the **why**, the next step is to determine **how** they can best build the infrastructure, tools, data systems, and capacity to incorporate that understanding and knowledge into their current business practices. Recognizing that businesses may be at different stages of their sustainability journey, and there-fore different levels of maturity in understanding and embedding sustainability into business practices and activities, maturity models are being developed and applied.

A maturity model or "matrix" is designed to define maturity levels for specific management practices. The columns represent increasing performance, capability, organizational structure, and/or stability of programs and processes (e.g., from ad hoc, essential, or initiating to transforming or culturally embedded), and the rows represent specific management practices (e.g., prioritizing risks and opportunities, or product transparency and traceability) that an organization should have in place to fully embed sustainability into its organizational culture. Figure 5.6 illustrates one example of a maturity-model framework for sustainability.

When determining which practices to pursue and how quickly, each organization must consider its individual circumstances and context, including its size, resources, strategy, priorities, and the estimated business value. Not all organizations will want to move across the model to the highest level of maturity for every practice. The key is for organizations to optimize their systems and programs based on their unique situation and goals.

Once an organization decides where to focus its efforts, it can use a maturity model to inform how to improve within each focus area. For example, let's say an organization is starting at an "ad hoc" level in the area of supply-chain management, and they have set a goal to progress toward the "transforming" level. While trans-forming is the ultimate goal, the organization will need to develop an action plan to get there. In some cases, the plan could involve drastic changes (e.g., business model

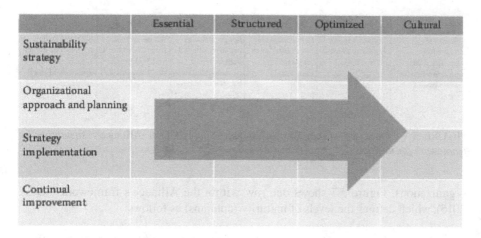

	Essential	Structured	Optimized	Cultural
Sustainability strategy				
Organizational approach and planning				
Strategy implementation				
Continual improvement				

FIGURE 5.6 Example maturity model from PSRT.

change) that would allow the organization to "jump" directly to the transforming level. However, for the majority of companies, more gradual, step-by-step changes are required to progress through the various maturity levels. Many organizations are adopting maturity model frameworks to embed sustainability into their organizational cultures.

Maturity models can be developed for various applications and can be broad (i.e., covering practices relevant to all organizations) or tailored to a specific sector or focus area (e.g., supply-chain management). The language used to define maturity levels differs and depends on the purpose and aim of the model. Although the titles of the levels vary, most maturity models use four or five levels of maturity.

Our case studies below illustrate how collaborative efforts among companies have developed and successfully applied maturity models to understand best practice, improve performance, and build more resilient organizations.

ELECTRIC UTILITY INDUSTRY SUSTAINABLE SUPPLY-CHAIN ALLIANCE (SSCA)

As highlighted in the collaboration section above, the Electric Utility Industry SSCA (the Alliance) uses custom tools and systems, including their sustainability framework, to help improve the supply-chain practices at each of the member utilities. The framework document provides an overview of the main elements or practices that are considered core to strong environmental sustainability management. The information is presented in a table that provides a means to gauge the level of maturity of corporate and supplier environmental sustainability programs and to help shape continuous improvement goals for organizations seeking to advance their programs. The practices within the table, which are each defined by five maturity levels (moving from Initiating to Progressing to Optimizing to Leading and finally, to Transforming), were compiled by sustainability and supply-chain subject-matter experts from the Alliance member organizations and other professional

	Initiating	Progressing	Optimizing	Leading	Transforming
Scope of suppliers included in environmental assessment	Scope includes suppliers at an ad hoc basis.	Scope includes select top tier or strategic suppliers.	Scope includes majority of tier 1 suppliers (e.g., those represented by 80% or more of spend).	Scope includes majority of tier 1 and high impact tier 2 suppliers.	Scope includes majority of tier 1, high impact tier 2 and critical tier 3 suppliers.

FIGURE 5.7 One row within the Sustainable Supply-Chain Alliance's *sustainability framework for utilities.*

organizations. Figure 5.7 shows one row within the Alliance's framework (SSCA 2016), which defines the levels of maturity (columns) as follows:

- *Level 1: Initiating*: Organization has some awareness of sustainability with ad hoc activities.
- *Level 2: Progressing*: Organization has a systematic structure and formal processes.
- *Level 3: Optimizing*: Organization is advancing internally with continuous improvement.
- *Level 4: Leading*: Organization has embedded sustainability practices and is recognized for some leading industry practices.
- *Level 5: Transforming*: Organization has developed and implemented innovative practices that transform industry sustainability expectations.

Alliance members have benchmarked themselves against the framework (i.e., identified at which level they are performing for each practice) and have set goals to continuously improve in multiple areas. The framework has also helped to inform the development of voluntary standards for the Alliance that address gaps and improvement opportunities identified by members and encourage best-practice sharing between members.

PRODUCT SUSTAINABILITY ROUNDTABLE (PSRT)

The Product Sustainability Roundtable, a global consortium of product sustainability leaders that engage in cross-value chain and cross-industry collaboration, has developed a maturity model or "Sustainability Leadership Framework" that can be used by any organization to embed sustainability into the organization. The Sustainability Leadership Framework has four levels of maturity including:

Level 1: Essential: Organization has basic elements needed to meet requirements and minimize risk. Responsibility is distributed to individuals, and a lack of programmatic processes means that practices are often ad hoc and inconsistently applied across the organization.

Level 2: Structured: Structured programs with common practices and procedures allow sustainability strategy to be implemented across the organization in a coordinated and consistent way. Responsibility is centralized into one or several dedicated experts, resulting in more efficient systems.

Level 3: Optimized: Systematic, organization-wide activities and tools are integrated into existing business processes and functions (e.g., product development and supply-chain management). Responsibility is shared among managers working together on implementation. Data analysis is used to optimize sustainability processes and performance.

Level 4: Culturally embedded: Sustainability is integrated into the business strategy and culture and is considered in all business decisions and activities. Responsibility resides with top management and involves taking a life-cycle perspective and collaborating with value chain partners. Proactive methods are used to predict risks and opportunities and capitalize on business value (e.g., opportunities for innovation, competitive advantage, and brand enhancement).

As with other maturity models, the rows within the Sustainability Leadership Framework represent specific management practices (e.g., stakeholder engagement, planning for uncertainty, and value chain innovation). This framework guides users through four levels of maturity for each of the specific management practices that are highlighted.

The Sustainability Leadership Framework is used to accelerate and scale PSRT members' product sustainability efforts by:

1. Defining key management practices required to develop a leading, culturally embedded organization.
2. Enabling self-assessment, identification of improvement opportunities, benchmarking against other member organizations and development of improvement goals.
3. Identifying common challenges across the membership.
4. Identifying best practices and resources that will enable continuous improvement within the framework.
5. Providing a structure to measure performance and progress against goals over time.

All organizations, regardless of where they are in their sustainability journey, can benefit from maturity models that define a path forward to building more resilient, agile organizations and ensuring continuous improvement over time.

ENGAGING THE RIGHT RESOURCES TO DO THE JOB-MANAGED SERVICES

As companies look to downsize or "right size" as has it been sometimes called, a question that often surfaces is, **how** we will continue to meet the demands of the legal requirements and market demands. Do I build the capacity internally or buy externally? Obviously, the answer is not easy. Perry (1993) outlined a simple decision logic that compared "proprietary and generic capabilities" with "is the work to be done value added or essential support work." For example, if the work is value-added support work but of a generic capacity, then the guidance was to develop ongoing access to the best capacity possible. However, if the work is value-added but proprietary in nature, then the guidance is to develop the best internal capacity possible. The authors have experienced this type of logic being applied within the product sustainability space. The following discussion provides insights into the rationale and lesson learned.

Ensuring compliance with material-related regulations, such as the EU Restriction of Hazardous Substances (RoHS) Directive and the EU REACH regulation, and driving toward greener products, requires a capable and efficient process to request and validate compliance data from the supply chain. The recognized best practice is to require suppliers to report full material disclosure (FMD) data (e.g., all substances present in each homogeneous material) for the components and materials they provide. Increasingly, companies are also requiring third-party laboratory analytical reports which quantify concentrations of analytes to verify compliance with EU RoHS and voluntary standards such as Low Halogen. Given the large number of components and materials that may be present in a product, this reporting process is highly transactional and potentially onerous from the standpoints of keeping up with the need for FMD data as new products are launched, validating that the data collected is complete and accurate and ensuring that the data collection process is conducted at the lowest cost of ownership.

Increasingly, this work is viewed as value-added but not proprietary in nature, and companies are looking at the opportunity to buy capacity externally as a managed service. This approach enables focusing corporate resources on strategic issues versus transactional work, at a lower cost than may be achievable via dedicated resources. Key considerations to ensure that the managed service meets corporate requirements for product compliance and development of greener products include:

- Standardization of supply-chain data requests around FMD data collection utilizing data exchange models such as IPC 1752A (Materials Declaration Management Standard. Latest revision: February 2014).
- Supplier on-boarding, training, and capacity building to ensure suppliers have the understanding and information needed to effectively respond to data requests.
- Scalable workflows and supporting software to accommodate evolving requirements related to material disclosure, conflict minerals, and supply-chain sustainability data.
- Integration with the customer's engineering systems and compliance system of record for compliance data. Alternatively, for customers who do not have an existing compliance system of record, establishment of a cloud-based system of record.
- Leverage off-shore technical resources for day-to-day transactional work to reduce the cost of the managed service and to ensure resources are available in the time zones where suppliers are located.

- Preparation of internal management reports as well as customer and regulatory reports.
- Continuous improvement of business processes and software automation to flatten and reduce overall cost of ownership.

A managed service should be set up under a Service Level Agreement (SLA) which articulates the services, technology, and key performance indicators which will be used to evaluate the managed service on an ongoing basis. An SLA is critical to ensuring that key parameters are understood and agreed to including that the desired outcomes of the managed service are achieved, which typically include cost reduction, improved supplier engagement, data quality improvement, support for metrics and reporting, compliance assurance, and reduction of product life-cycle impacts.

IN THE END, IT IS ALL ABOUT CHANGE MANAGEMENT – EMBEDDING A PRODUCT SUSTAINABILITY TOOL – A CHANGE MANAGEMENT SUCCESS

As the field of sustainability performance advances, decision-support tools allow more and more useful data to become available. However, **even if one has useful tools, they are often not adopted by companies – not for an obvious reason**, but often it is because "it was not invented here or with me as a key stakeholder" or "I have done it this way for years and I do not want to change – why should I?" and "how will it allow me to get my job done faster and easier" are often the responses given. Johnson & Johnson developed its Earthwards® approach, an excellent tool/system to inform innovation teams about hot-spots or priority areas for targets for improvement. Moreover, Johnson & Johnson has made it a requirement for all products going through the NPD process to use the Earthwards® approach. What were the actions, success factors, and approaches to overcome obstacles that Johnson & Johnson undertook to make Earthwards® an integral part of its business practice? These are examined in the case study below through a series of questions that were asked of the developer of the Earthwards® process, Al Iannuzzi.

JOHNSON & JOHNSON EXAMPLE

What Was Done to Embed Earthwards® into New Product Development Processes?

First, we developed a product stewardship requirement to be a separate chapter in our Environment, Health & Safety (EHS) & Sustainability Standards. The Standards are business requirements which all operations must follow. Part of the requirements were that Earthwards® checks had to be integrated into the NPD processes of the various business units. The implementation of these checks varies based on the different processes to bring new products to market.

What Were the Obstacles, and How Did You Overcome Them?

The main issue with embedding sustainability into NPD is the concern with slowing down the development process. The key is to **make the checks easy, fast, and meaningful**. Getting the help of product stewardship professionals to give advice and insights to product development teams was beneficial. Web enabling our scorecard and simplifying the documentation required was important so that we didn't turn off brand teams that were applying for Earthwards® recognition.

What Were the Success Factors?

There were several key success factors; for starters, getting buy-in from key R&D management to make meaningful checks, and embedding key questions into the NPD process was critical. **Connecting Earthwards® to our sustainability goals** was helpful to focus business units' efforts since we have a very robust reporting system for our goals (2020 goals are for 20% of JNJ revenue from Earthwards® products). Soliciting customer insight and demand for greener products via our standard voice of the customer business processes and sharing this with management helped bolster the business case. Doing internal and external marketing of our Earthwards® recognized products was critical – making people aware of new products that met our criteria that were being recognized helped to spread the word and create demand. Having our key customers aware of our approach and the things we are doing to meet their demands helped our management to see the benefit to their business and their marketing programs, for example, highlighting Earthwards® products to customers during the request-for-proposal processes and in one-on-one meetings with purchasing managers.

If You Were Going to Do This Again, What Would You Do Differently?

We learned a lot along the way: simplifying and web-enabling tools, tying Earthwards® to processes that were being used by business units, expanding our criteria to address business and customer needs, such as the removal of materials of concern and social issues all built momentum for Earthwards®. We would have liked to have made these enhancements faster and to have anticipated some of these things. Also, we now position **Earthwards® as an innovation tool.** Becoming more aware of the potential of sustainability as an innovation driver would have been a good thing to do earlier on in the process. If we had worked earlier with our procurement R&D organizations, it would have helped us to develop even more Earthwards® products.

The Ultimate Success Factor – Linking Product Sustainability and Business Value

Historically, there has been a perceived disconnect between improved product sustainability and additional value for business. The traditional rationale is that sustainability attributes aren't valued by the marketplace. However, there are also often quoted examples of very successful product sustainability programs that continue to grow and prosper more than 10 years after introduction. GE's highly successful Ecomagination program has gene rated "$160 billion in Ecomagination product

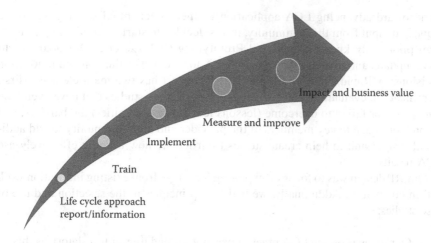

Impact and business value

Measure and improve

Implement

Train

Life cycle approach
report/information

FIGURE 5.8 Outcomes to impacts. (Adapted from Global Environment Facility (GEF). 2009. *The ROtl Handbook: Towards Enhancing the Impacts of Environmental Projects – Methodological Pape*r # 2. www.gefeo.org.)

revenue" and resulted in a "32% reduction in GE's greenhouse gas (GHG) emissions from the 2004 baseline (GE 2017)." Johnson & Johnson's Earthwards program has recognized products that are valued at $9 billion as of 2015. Why do these programs continue and thrive if such a disconnect exists?

Part of the answer may lie in the very nature of product sustainability tools and practices.

There is a tendency within the sustainability space to focus on activities – number of LCAs, sustainability reporting, development of standards, scale of product reviews, etc. – rather than on **outcomes**, absolute impact reductions, displacement of less sustainable alternatives, etc. To drive the latter requires a commitment to driving activities through the additional phases of implementation, measurement, and improvement. A telling example is the use of an LCA. An LCA that is completed and put on a shelf does nothing to address value generation or impact reduction. By contrast, an LCA that initiates a process of engagement with product development teams, identification of opportunities and goals, measurement of improvements, and linkages to market demand is dramatically more valuable to an organization (Figure 5.8). In other words, many product sustainability efforts never get beyond assessment to the point of actual improvement – improvements that can matter to customers.

COLLABORATIVE PROJECT CREATED INSIGHTS TO MOVE LIFE-CYCLE INFORMATION TO THE BEGINNING OF INNOVATION

A multiphase project, jointly supported by Society of Environmental Toxicology and Chemistry (SETAC), American Center for Life-Cycle Assessment (ACLCA) and Forum for Sustainability for Life-Cycle Innovation (FSLCI), was developed to work toward advancing LCA application further. After initial scoping and some requested input from the community, it was decided to start by conducting a short-term pilot study known as a Rapid Prototyping (RP) exercise. Our goal was to drive, update, and increase the value of LCA for product evaluation and innovation decisions: A Rapid Prototype Pilot. This RP effort has two main elements: first, identification, evaluation, and critical analyses of cases studies that have overcome (successes) or failed to overcome (lessons learned opportunities) the barriers; and second, outreach to key members of the provider and user community to add additional insight and to help create stories to articulate how to more effectively use LCA results.

Our RP design was to focus on exploring learnings from existing innovation work within companies. Additionally, we had four principles in the selection and use of case studies:

1. Our focus is on the LCA practitioner (not the end user or translator), as this is the role, we are in the best position to intervene with/influence the system they work within.
2. Prioritize the cases in our collection where one of us was the LCA practitioner so that we can rely on firsthand knowledge.
3. Next priority are cases where one of us has a good relationship with the practitioner (candor will be key).
4. Do not analyze cases with more than one degree of separation between this team and the practitioner.

The case studies included Bemis Plastic Shopping Cart – BASF, Braskem Disposable Cups, Epoxy Kitchen Sinks, Integrating LCA-Based Carbon Performance at Interface, and Steelcase Think Chair Refresh. For each case study the following criteria were evaluated – specific successful outcome related to LCA, why is this remarkable, success enabler and barrier overcome, timing or project stage, and potential strategy to replicate.

It is important to acknowledge and thank the excellent collaboration among the three organizations, ACLCA, FSLCI, and SETAC who sponsored the project. Bruce Vigon (Breveja, USA) and I were the co-chairs of the multiphase project. Jeff Zeman (True North Collective, USA) and Jeremy Faludi (TU Delft Sustainable Design Engineering, Netherland) were the co-chairs for this pilot study. The rest of the team consisted of Lina Cowen (Synapse, USA), David Evers (Consultant, USA), Dustin Heiler (Steelcase, USA), Stefanie Koehler (Green Crane Agency, USA), David Meyer (Environmental Protection Agency, USA), and Barbara Nebel (thinkstep-anz, New Zealand). Many thanks to all the team participants and the support of the companies.

One of the initial realizations in our planning was it is hard to develop a plan to bring LCA to the beginning of innovation if we do not have clarity in where you want to go. The following was the team's vision of success.

- LCA providers consistently **deliver simple, relevant, and actionable feedback** to designers and decision-makers.

- **"Life cycle literacy" is built into students' curriculum** across a variety of disciples (e.g. engineering, MBA).
- **LCA-sourced information is served in a custom manner** that is primed for case-specific understanding and application.
- The **role of LCA providers is elevated** to the point that they are sought out to provide a must-have component to balanced decision-making.

ULTIMATE SUCCESS

LCA is one of the regular tools used within innovation (to inform design decisions to create more sustainable products that achieve positive impacts. These conversations and outcomes were helpful. It is hard to figure out what policy, practices, and tools are useful, if you do not know where you are going. As shared in Golden Rule 6, If you do not know where you are going, any road tool will get you there (according to Alice in Wonderland – Cheshire Cat's comment to Alice).

Lessons Learned

The outcome of the collaborative project revealed the following six learnings:

1. LCA information is crucial at the beginning of innovation to set priorities for innovation. It is also valuable throughout the design/innovation process to compare options.
2. Life-cycle perspectives are used in different ways throughout the design/ innovation process within accepted stage gates.
   a. At the earliest stage, life cycle thinking (LCT) helps us prioritize hot spots and find opportunities.
   b. As the process continues, it helps us compare alternative options against each other and/or against specified goals.
   c. At launch approval, we can use LCA to "report-out" – have we created a more sustainable choice?
3. As LCA practitioners, it's our calling to create tools and processes suited to these stages.
4. Always use LCT, sometimes do LCA – Do LCA when detail and precision will help make a point or inform a decision. Applying LCT right at the beginning of product development helps to get a full understanding of the entire supply chain, which helps to uncover other potential risks along the supply chain, such as modern slavery, critical materials, and potential supply risks due to geopolitical circumstances.
5. We can't do this alone – integration with design/innovation process requires buy-in and assistance from those we wish to influence, e.g., innovation managers, chief technology officer, material experts, process owners, supply-chain leaders.
6. Get buy-in early and often from senior decision-makers. Operations and engineering (often the data providers) will be more supportive and open to the process.

Initial possible actions to be taken to move LCT/LCA to the beginning of innovation are described in Table 5.1, based upon four levels of sustainability leadership.

Several broader lessons were realized. They are:

1. Sustainability associated with your products is becoming a power driver to change companies' strategies, brand/business practices, and financial performance.
2. If you have not developed an understanding and have a plan to embed sustainability into your core businesses (e.g., innovation, and procurement), you may be left behind. Obviously, as we will share latter in the Golden Rules, success is best achieved by clear policies, metrics, and actions which are part of core business practices – i.e., governance.

TABLE 5.1

Actions to Move LCT/LCA to the Beginning of Innovation

LCA/LCT Moved to the Beginning and throughout Innovation: Initial Actions Based upon Your Strategy/Ambition Level

	Level 1: Initiating	Level 2: Structured	Level 3: Optimized	Level 4: Culturally Embedded
General	• *Basic elements to meet requirements and minimize risk.* • *Responsibility distributed to individuals.* • *Lack of programmatic processes means that practices are often ad hoc and inconsistently applied across the organization.*	• *Structured programs with common practices and procedures result in coordinated implementation.* • *Responsibility is centralized into one or several dedicated experts, resulting in more efficient systems.*	• *Systematic organization-wide activities and tools integrated into existing business processes and functions.* • *Responsibility shared among managers working together on implementation.* • *Data and analysis used to optimize processes and performance.*	• *Sustainability integrated into business strategy and culture and considered in all business decisions and activities.* • *Responsibility resides with top management and involves taking life-cycle perspective and collaborating with value chain partners.* • *Proactive methods used to predict risks and opportunities and capitalize on business value.*

(Continued)

TABLE 5.1 (*Continued*)
Actions to Move LCT/LCA to the Beginning of Innovation

LCA/LCT Moved to the Beginning and throughout Innovation: Initial Actions Based upon Your Strategy/Ambition Level

	Level 1: Initiating	Level 2: Structured	Level 3: Optimized	Level 4: Culturally Embedded
Move LCA/LCT to the beginning	• Reach out to customers, peers, and thought leaders to explore what, why, and how to embed LCA/LCT into innovation. • Reach out internally with sales, marketing, and innovation to understand what they are hearing from the marketplace. • Develop a plan to pilot and test consideration of LCT/LCA. • Provide awareness building with EHS and innovation teams.	• LCT/LCA Is being piloted within one of your brands/business groups to understand risks/opportunities related to the analysis of material or design options. • DfS role is established. • Awareness building, benchmarking continues building off of results of pilots. • Plans with business case formulated to further embed LCT/LCA into innovation.	• Life-cycle perspectives are used in different ways throughout the design/innovation process within accepted stage gates. • At the earliest stage life cycle thinking helps us prioritize hotspots and find opportunities • As the process continues it helps us compare alternative options against each other and/or against specified goals. • At launch approval we can use LCA to "report-out" – have we created a more sustainable choice? • Communicate knowledge – not data.	• Sustainable innovation is an agenda item with executive management. • Enterprise and business units/brand goals are set for sustainable innovation. • Progress on sustainable innovative is a key component within annual reporting and shared on website(s).

3. Always use LCT, sometimes do LCA. Do LCA when detail and precision will help make a point or a decision.
4. Translate LCA results into your audience's language – engage your audience right from the start (Golden Rule 4).
5. Resources exist, build capacity to use LCA studies to drive innovation, identify trade-offs, and ensure unaccepted consequences do not arise (Golden Rule 10).
6. The best use of LCA is to identify, understand, and navigate trade-offs. Circularity vs. carbon footprint is one of the biggest trade-offs of our decade.

The rise of sustainability awareness in the market has created substantial drivers in the consumer market, which in many instances may have not existed in the very recent past. The key realization to keep in mind is that every sustainability improvement does not resonate equally in every market. A recyclability claim may be very powerful for a packaging component that has very little perceived value to the end consumer, who otherwise is faced with no other option than landfill. Conversely, recyclability for a longer-lived energy using product may be of much lower importance than energy efficiency that contributes to lower cost to operate. Mapping received value related to sustainability onto specific market drivers can result in clear messages that resonate with customers.

In business-to-business markets, the realized value is usually easier to identify and explain. Most companies of all sizes are actively striving to improve their own sustainability performance, many of which are publicly reporting on their objectives and performance. These sustainability frameworks can be seen as a roadmap for suppliers to navigate, crafting sales messages to their customers of how their products can help them meet their goals. Some companies have built sophisticated programs to support this effort. A common theme among those programs is to combine resources to support both the R&D teams in reducing impacts, as well as the sales arm to effectively communicate the financial and sustainability benefits to customers. BASF provides an excellent example in their Sustainable Solution Steering program, which identifies and encourages development of "Accelerator" products that not only meet all sustainability requirements but also provide demonstrable benefit to customers, see the earlier description of BASF enhancement of the Portfolio sustainabilty assessment approach. Known as "Triple S," BASF quotes more than $16 billion of Accelerator solutions since the outset of the program, with more than 60% of their R&D spending being related to improving sustainability performance (BASF 2017).

To fully realize the potential for sustainability requires harnessing the systems, processes, and brainpower within one's organization. This in turn typically requires translating sustainability "lingo" and desired outcomes into the traditional business metrics of revenue, costs, brand, and risk. It is rare for any one activity to address all four elements equally; therefore, to truly optimize outcomes often requires focusing on one or two areas as priorities. For example, a company that pursues a greener product strategy is likely to benefit from engaging their marketing team to help communicate the performance and impact reduction benefits from their products in order, internally and externally, to realize results. The more that these contributions can be quantified and forecasted the better, but experience has shown that even operating within this framework can be powerful in aligning resources and realizing results (as shared in Golden Rule 4).

GOLDEN RULES FOR A SUCCESSFUL PATHWAY TO MORE SUSTAINABLE PRODUCTS

Products/technologies are designed, manufactured, used, and managed at the end of life in ways to create reduced environmental and social impacts, considering a circular economy. Unintended consequences are identified and considered in decision-making. These innovative products create business and societal value. All actors over product/resource/packaging life cycles have a role to contribute to creating societal and business value. New business models surface. The life-cycle community provides relevant information to inform decision-makers. Innovation teams play key roles to embed sustainability metrics into their stage gate processes.

These are just some of the changes surfacing and being incorporated more and more into core business operations. Successes are driven by several commitments; for example, top management has a strong commitment to establish metrics, which are reviewed and evaluated regularly. The Golden Rules have been established and evolved based upon years working with companies around the world to help them understand what sustainability is, whether and how it applies to them, clarity on what they can do about it, and actions that should be taken at the enterprise, business unit, and product levels. They have been called our guiding rules to realize the value of applying life-cycle information. When you are reading these ten Golden Rules, identify actions you can take on Monday morning that will move you and your organization toward more sustainable products. Remember, these Golden Rules reflect a management system's perspective which creates the demand with the organization for more sustainable products. We understand what makes LCA successful and have learned the keys to influence how we sustainably design and develop new products, technologies, and services for the 21st century.

Companies have been working to develop efficiency and effective strategies and tactics to embed sustainability into their operations for many years. Based upon these experiences, ten Golden Rules have been developed and applied. In this section, they will be described and examples from their applications will be presented. **The Ten Golden Rules are:**

1. There are no green (sustainable) products/packaging – only "greener" or "more sustainable".
2. LCA provides solid understanding of impacts over the entire life cycle.
3. LCA information is essential but not sufficient.
4. Use the language of your audience.
5. If you don't know your destination, any tool will get you there.
6. Without a seat, three legs of a stool are useless.
7. Meet them where they are.
8. Focus is on positive impacts – not just study results.
9. It is all about change management.
10. There is a growing life-cycle community – all around the world – they are a resource – access them.

Building off the publication *Ten golden rules for applying Life-Cycle Information* (Fava, J, 2021), this section expands upon that information. Let's explore each of these Golden Rules further.

> *Whenever I speak about greener products, there are two things I usually say:*
>
> 1. *There is no such thing as a green product.*
> 2. *What good is a greener product if no one knows about it?*
>
> —*Al Iannuzzi*

GOLDEN RULE 1: THERE ARE NO GREEN (SUSTAINABLE) PRODUCTS/ PACKAGING – ONLY "GREENER" OR "MORE SUSTAINABLE"

The first Golden Rule is "All products/packaging have some type of impact – there are no green products/packaging – only 'greener." Full credit for this Golden Rule is given to Al Iannuzzi, the author of this book – *Greener Products*, the making and marketing of sustainable brands. He is quite clear when he said: "*Whenever I speak about greener products, there are two things I usually say:*

1. *There is no such thing as a green product.*
2. *What good is a greener product if no one knows about it?*

The reason for these assertions is that life-cycle assessments have shown that every product has impacts, from raw materials to transportation, manufacturing, customer use, and end of life. Every product can be improved in some way, which is why I use the term Greener."

All products use materials, resources (e.g., water, metals, chemicals), those are transported from a supplier (s) to a manufacturer, who use energy to transform those materials, parts to a product, which is then transported to a customer and/or consumer, who then uses the product and disposes of the unused product or package, either in a landfill, or recycled, or another end of management solution. All of this uses energy, materials, water, and releases emissions into the air, water and onto land, and wastes at the various stages of the product's life cycle. Use of LCA allows a company to know where those resources and emissions occur to identify hot spots for which can be targeted to reduce or minimize. **Thus, the term Greener – not Green**.

The Coca-Cola Company recognized that all the materials they used as a container for their beverage had environmental impacts. They challenged the beverage container sector to develop a packaging material with reduced impact. The Aluminum industry developed recycled aluminum that could be used for the Coca-Cola beverage container. The recycled beverage container helped establish the early recycling industry. A key to Coca-Cola's decision is that they recognized that all materials had impact and had challenged their suppliers to reduce their impact. Clearly, they had recognized the need for a greener beverage container.

It has been a pleasure to have had the opportunity to work with Al Iannuzzi for many years in both supporting his efforts within his company as well as collaborating with him on these three editions.

When an organization makes a claim that they have green products, it would be good to check with them about their claim, what is the basis for that claim, and how do they justify the statement(s).

GOLDEN RULE 2: LCA PROVIDES UNDERSTANDING OF IMPACTS OVER THE ENTIRE LIFE CYCLE

There are three elements in this Golden Rule – (1) *What is LCA?* (2) *What are the life cycle stages?* (3) *What are the impacts?* The following sections address each separately.

What is an LCA? LCA provides organizations with information about the environmental and social impacts of a product or service throughout its entire life cycle. It is a systems analysis tool to inform environmental decision-making for sustainable development. LCA provides:

- Solid understanding of impacts over the entire life cycle.
- A means to identify trade-offs based upon identified hot spots for targets for improvement – e.g.,
  - Presentation of trade-offs among multiple human health and environmental issues
  - Presentation of trade-offs between different elements in the value chain, such as life-cycle stages, economic actors
- The ability to understand and communicate counterintuitive environmental benefits that would not have been revealed through any other approach than LCA.
- A whole system consideration – avoid shifting the burden – being able to understand and avoid unintended consequences.

An example is a shoe manufacturer interested in whether she should move the manufacturing of the final shoes back from Portugal to the USA. She was concerned about the logistics and transportation greenhouse emissions linked to shipping the leather from the USA to Portugal and subsequently shipping the final goods back to the USA, which is her primary market. A streamlined life-cycle study revealed that the greenhouse gas emissions related to growing the cattle and processing the leather were more of an impact than the transportation component. Based on that information, she decided to keep the shoe manufacturing in Portugal and then work with the cattle sector to address leather production. Simple example, but it illustrated how LC information can help inform decision-making.

Within LCA applications, there is both the supplier of life-cycle information and the user of that information. The supplier of the life-cycle information community understands the expectations/needs of the decision-makers, so their LC information is relevant and understandable. Decision-makers understand and know how to incorporate the life-cycle information into their decision-making processes.

As we have shared, the modern-day LCA was developed initially through the SETAC efforts and workshops. Then we had the ISO 14040 series of LCA standards. We have organizations and governments around the world continuing to advance and apply LCA studies to inform decision-making.

I have been asked if you would change anything related to the early development of LCA. Overall, the early work by SETAC globally is still the foundation of today's LCA methodology. Obviously improved and enhanced by ISO and other organizations and governments, the four phases and key components still exist. There is one area, when I look back on it, I would have voted against the change from improvement assessment to interpretation. At the SETAC workshop in 1990, the attendees agreed that the LCA method should go beyond the energy and material mass balance methods that had been used for a few years to include an assessment of the materials and emissions – this impact assessment was added. Additionally, the LCA results should drive change – i.e., improvements to the product system. Coming out of the SETAC efforts, LCA consisted of Goal and Scope Definition, Inventory Analysis, Impact Assessment, and Improvement. During the ISO standard development efforts in the early 1990s, the countries changed Improvement Assessment to Interpretation. The rationale was that Improvement is only one outcome and use of the LCA results. That is one area I would take back. It is also why my focus has been and continues to be to **embed LCT/LCA into innovation processes to make product improvements**.

What are the life-cycle stages? Figure 5.9 illustrates typical life-cycle stages from extraction of raw materials to end-of-life management. Initially it was a linear life cycle – from cradle to grave or cradle. With the understanding of where product's impacts occur, and solutions to minimize or eliminate those impacts, a circular design is rapidly evolving to be foundational in innovation processes.

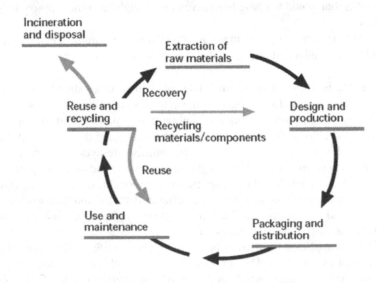

FIGURE 5.9 Typical life-cycle stages from extraction of raw materials to end-of-life management.

The third component is *what are the impacts*. LCA contains many impact parameters. Explicit in the definition of an LCA, it includes the multiple life-cycle stages, as well as multiple impacts. Examples of typical impacts include, but not limited to Acidification Potential (AP), Ecotoxicity Potential (ETP), Eutrophication Potential (EP), Global Warming Potential (GWP) (Also: Climate Change), Human Toxicity Cancer Potential (HTCP) (Also: Human Health Cancer), Human Toxicity Non-Cancer Potential (HTNCP) (Also: Human Health Non-Cancer), Human Health Criteria Air Potential (HHCAP) (Also: Human Health Particulates), Stratospheric Ozone Depletion Potential (OPD) (Also: Ozone Layer Depletion), and Smog Creation Potential (SCP) (Also: Photochemical Ozone Creation).

Initially, the focus was on environmental impacts. Over the last few years, there have been advancements to add social impacts into the assessments (UNEP LCI 2022.

For a study to be an LCA, it should include multiple impacts, not just one. The value of LCA results, as noted, is to help identify hot spots and trade-offs. With a focus on only one impact (e.g., GHG emissions or recycling content), an LCA study will identify hot spots along the various life-cycle stages. It would not be able to provide any insights into possible unintended consequences of deciding to address one impact while, doing so, create another.

An example (many thanks to Gerald Rebitzer – AMCOR – who shared this example during a PSRT meeting) is if someone is interested in buying a car and wanted it to have a reduced GHG emissions associated with its use, the rapidly expanding electric vehicles would be an obvious choice. If on the other hand, the person wants to have a vehicle that could be highly recycled, then the 1970 thunderbird made from steel would be perfect. Obviously, that comparison is not realistic. But it does illustrate that designing a vehicle to maximize its impact during the use stage creates one design. Maximizing a product to be able to recycle AND reuse the materials would create another design. The underlying premise of sustainable innovation is designing a product which addresses multiple hot spots. Some can be part of the ongoing operation of the vehicle and others can be a partnership with parties who can reduce those impacts later in the life span of the vehicle (e.g., reuse/recycling sector). This theme will continue throughout this chapter.

Golden Rule 2 – To design more sustainable products, you must first understand the full environmental and social impacts over the entire life cycle.

GOLDEN RULE 3: LCA INFORMATION IS ESSENTIAL BUT NOT SUFFICIENT

In decision-making, **LCA information is essential but not sufficient**. LCA is a tool within a broader array of tools to understand the full range of environment and social issues associa21hunted with a product/packaging system. Many other factors, in addition to LCA results, will be considered in the decision-making, including social aspects, technical feasibility, costs, and supply-chain risks. An example is highlighted in the UNEP report based upon efforts by the appliance industry (AHAM 2012). In this example by the American Home Appliance Manufacturers (AHAM), LCA results were essential but to reach a positive outcome, multi-stakeholder concerns,

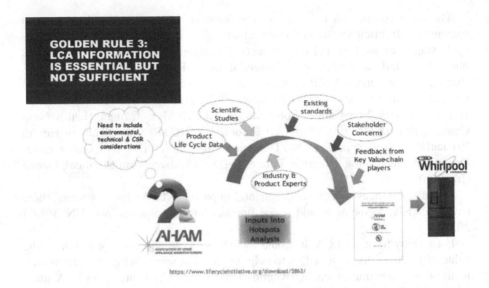

FIGURE 5.10   LC-based product life-cycle data is an integral role to play. It is clearly essential but not sufficient.

value chain insights, experts on specific impacts, existing standards and concerns are just a few of the additional information needed. In the previous edition of this book, the AHAM example is described in detail. It is a good illustration of this Golden Rule, so I wanted to summarize again. The goal of the AHAM focus was to develop design guidelines based upon hot spots analysis that could be used by the appliance companies to develop the next generation of appliances (Figure 5.10).

The hot spots analysis framework used input from environmental, technical, and CSR considerations, coupled with product life-cycle data, scientific studies, standards, stakeholder concerns, feedback from key value chain players, and industry and product experts. Clearly LCA data is a critical part of the analysis but is not the only source of data and information.

This example used a systematic process to identify the hot spots, which together with key product and stakeholders developed a set of targets that could be used by the designer to develop the next generation of appliances – greener than the previous generation.

A multi-stakeholder group was used early and throughout the process to reach alignment of the criteria and the performance expectations. The multiple stakeholder group was called the stakeholder advisory council (SAC) and consisted of professionals from NGO, government, business, and technical communities. As we will discuss later, this is an example of where there is the need for both tools and information but also the perspectives from a user's and broader array of stakeholders.

GOLDEN RULE 4: SPEAK THE LANGUAGE OF YOUR AUDIENCE

As we have indicated throughout this chapter, **being able to speak in the language of the receiver is critical** and can influence your abilities for them to fully

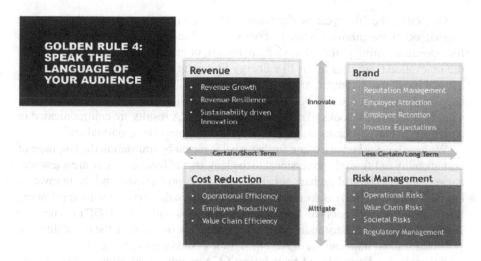

FIGURE 5.11 Core business values achieved with the use of sustainable information.

understand and act on your message/request. A key outcome of several years of working with companies has been a very simple 2 × 2 which we've taken to referring to as our "Rosetta Stone" because it helps us translate sustainability into the traditional sources of business value – Revenue, Brand, Cost, and Risk (Figure 5.11). This isn't to discount the other aspects of sustainability – the environmental and social – but rather we see the opportunity to fully engage businesses in these efforts as the best opportunity to accelerate the needed improvements in the others.

These four major values – grow revenue, enhance brand, reduce costs, and mitigate risks and the additional sub values for each – should become part of your language if you want to influence the decision-maker. We have described these in earlier versions of this chapter. Here we wanted to share thoughts on different ways of communicating.

First, business value is more than just tightening your belt. Companies must make an explicit choice regarding their sustainability strategy. Is the plan to only be in compliance with laws and regulations in jurisdictions in which they operate, with no additional efforts beyond compliance? Or, are there specific drivers in your market that demand response, but not an opportunity to differentiate yourself? Are those market drivers substantial enough that an engaged strategy will drive additional value related to brand enhancement or risk mitigation? Or, is there an opportunity to shape the future of your market, with commensurate additional revenue growth expectations attached? There are rational reasons for choosing any of these strategies, and there are organizational implementation ramifications that need to be considered as well. The bottom line is that cost savings is always a great business value, but it's a safe bet that there are competitors who are seeing more of an opportunity and moving beyond compliance and cost savings – e.g., risk mitigation, brand enhancement, and growing revenue. Perhaps they know something you don't?

Generally, the life-cycle professional uses language which is not the same or understood by the innovation teams. For example, life-cycle professionals use terms like global warming potential, toxicity potential, or eutrophication potential. While the innovation team uses terms like Energy Star rating, black or gray lists, durability, and biomaterial content, among their normal attributes (e.g., costs, reliability of suppliers). As we discussed in Golden Rule 5 to achieve success, one must have a view of what success looks like. Our vision is that LCA results are communicated in a manner that are understood and usable by the designers/decision-makers.

We have been asked to explain further what is meant by translating the language of sustainability or life-cycle professional to the language of business. Here are a few scenarios with what typical exchanges between the LCA professional and the innovation engineer should be (full disclosure, as a test of AI, given the increase in its application, these examples were adopted from several exchanges with CHAT GPT). They are consistent with my understanding and experiences, but to illustrate the capability and possible uses and implications of AI, I thought it would be good to share.

Scenario 1 – Examples of translating LCA results to attributes that can be used by innovation engineers. Approaches it from an impact-by-impact basis.

Material Selection

LCA results: The LCA professional LCA identifies that the production of a specific plastic used in a product has high greenhouse gas emissions.

Innovation attribute: The innovation engineers could select an alternative material with a lower carbon footprint or explore opportunities to use recycled or bio-based plastics.

Waste Management

LCA results: The end-of-life disposal of a product in landfills potentially has significant environmental impacts.

Innovation attribute: Innovation engineers can focus on designing products for recyclability, increasing product longevity, or facilitating easier disassembly for component reuse.

Supply-Chain Optimization

LCA results: Significant impacts are associated with a particular step in the supply chain.

Innovation attribute: Engineers can work closely with suppliers to adopt more sustainable practices or even redesign the product to avoid that step altogether.

Scenario 2 (More specific scenario) – A new electric vehicle (EV) model uses a battery that, while reducing fossil fuel consumption during its use phase, also poses a significant environmental burden at the end of its life due to landfill waste.

LC Professional Example

Our recent LCA reveals that while our new EV model substantially cuts down on fossil fuel use, the end-of-life management of its battery presents a significant environmental challenge. The materials and design make it potentially problematic for landfill disposal. We have a classic trade-off between reducing greenhouse gas emissions during use but increasing potential landfill waste at the product's end of life.

Innovation Engineer Responses (Examples)

Repurposing used batteries: Although a battery might no longer be efficient for vehicle use, it might still hold enough capacity for less demanding applications, like stationary energy storage. We can explore a secondary market for "spent" EV batteries.

Design for disassembly: Let's consider redesigning the battery to be more modular. If it's easier to disassemble, we can more efficiently recover valuable materials, reducing landfill waste.

Educational initiatives: We should educate our customers about the importance of responsible battery disposal and provide them with resources or incentives to return batteries at the end of their life.

Stakeholder engagement: Engaging with environmental NGOs or community organizations might provide us with insights and partnerships to tackle the battery disposal challenge effectively.

Collaboration with regulatory bodies: Let's work closely with local and national regulatory agencies. They may have, or may be in the process of developing, guidelines and infrastructure for EV battery disposal.

Continuous monitoring: Given the rapidly evolving nature of battery technology and recycling processes, we should continuously monitor advancements in the field to adapt our strategies accordingly.

In this second scenario, additional example responses were added, because it reflects a potentially actual situation battery manufacturers and users of those batteries (e.g., electric vehicle companies, and you and I consumers) are facing today.

Another example, shared previously, still conveys a clear outcome message and is relevant to this conversation. In a workshop in New Zealand, we used this business value framework in a series of interaction sessions with examples. Each attendee goes through their own examples and ideas. At the start of the workshop, I asked them to think of a project they had wanted to be funded within their own company that had not been funded. After the four-hour workshop, I asked them if they applied the language they just learned, do they think their project would have been funded – all 40 of them raised their hand and said yes – **speaking the language of the receiver improves your likelihood for success** (Fava 2021).

As you can read by the number of examples and stories, Golden Role 4 is key. Success requires the ability to translate the sustainability and life-cycle language into attributes that are understood and used by the innovation teams.

Golden Rule 5: If You Don't Your Destination, Any Tool Will Get You There

I always loved the movie Alice and Wonderland. One of my favorite characters was the Cheshire Cat. A line that resonated with me many years ago was, "If you don't know your destination, any road will get you there." Seems logical to adjust it to – *If you don't know your destination, any ~~road~~ tool will get you there.*

We have often worked with companies to help them develop tools. Occasionally, the work was within a technical group within the company – perhaps the EHS department. A design for environment tool or an LCA tool/study was conducted to help identify impacts or hot spots of the products. When the tool was developed or the study completed, if the users of the information, e.g., marketing or R&D were not engaged in the effort, the results were often viewed skeptically or fell on deaf ears by the potential users – wasted efforts. **Tools which do have value may lose their potential to be useful within a company if there is not an overall strategy and governance within the tool**. Golden Rule 6 addresses the importance of governance – without a seat, three legs of a stool are useless.

As we shared in the *What (vision), why and how* lesson learned, one must be clear about what success looks like. Hard to understand the WHY and develop the HOW if you do not know where you are going – what does it look like when you are there.

We provided an example application of understanding what success looks like in our 4th section on rapid prototype project. Specific examples also related to bringing the users of a life cycle or sustainability study into the goal and scope definition (the first phase of an LCA), so they can share their expected decision-making application of the results, for example, selecting material for use in the product. An example of initial focusing on the What or vision is with the Pilot Project that is described in Section 4. We used the ABCD framework to help us align on what success looks like and how to get there (adapted from Walsof Pty Ltd. (2003). AcdB™ Questions and Outputs. 2nd Road Pty Ltd.)

> A – *Current state*: *Where are we now? Assess the current situation and challenge.*
>
> B – *Desired state*: *Where do we want to go? Define the goal and desired outcome.*
>
> C – *Strategy development*: *What do we do to get there? Develop strategies and decide on actions to from A to B*
>
> D – *Action plan*: *How do we make it happen? –- Outline the steps and resources needed to implement the strategies.*

A final example – let's take an analogy/A builder who does not build the house with tools alone, she also builds the house with a design blueprint for the building – without a blueprint, a tool will not build a house. Thus, Golden Rule 5 – you need to know where you are going for you to select and use the right tool(s).

GOLDEN RULE 6: WITHOUT A SEAT, THREE LEGS OF A STOOL ARE USELESS

When sustainable development was introduced in the late 1980s, its focus was on three global priorities, Social, Environmental, and Economic. Examples of specific priorities for each are:

Social
Job creation
Local economic impacts

Business ethics
Diversity
Human Rights

Environmental

Products and operations
Resource efficiency
Zero waste
Climate change

Economic

Capital efficiency
Risk management
Margin improvement
Growth enhancement
Total shareholder return

Early in my career, it became obvious that when these priorities were not embedded into the normal operational or business practices, actual sustainability performance appeared to be ad hoc or inconsistent. For example, when a person took leadership to drive action and specific project(s) to use life cycle/sustainability metrics into a product design, improvements were realized for those projects. However, when that person took a new position or left the company, those actions were reduced or just did not happen. The business practice had not been embedded. We refer to **embedding into the business practice as part of the governance**.

We have often used a three-legged stool analogy, each leg of a stool representing sustainability performance from an environmental, social, and economic perspective (Figure 5.12). The companies making regular improvement are ones who

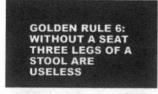

FIGURE 5.12 Three legs do not become a stool unless it has a seat.

also have a governance system which recognizes and incentivizes that sustainability performance.

We've worked with many of these leaders and the value of that governance is palpable. Interestingly, this is the value that external investors are looking for and are using sustainability performance metrics as proxies as they make investment decisions. Over the last 5–10 years there has been an explosion in the use of ESG (environment, social, and governance) metric as a guide to sustainability performance. Although an anti-ESG movement surfaced in early 2020s, there is a solid and strong global commitment by a wide diversity of countries, companies, governments, organizations, and particularly the younger generation is rising to become advocates for sustainability performance. Sustainability is a core market aspect for all organizations to manage, and those who do it well, do it in alignment with all the other aspects they manage. With a seat – i.e., the governance – the three legs (environmental, social and economic) become a very useful stool.

GOLDEN RULE 7: MEET THEM WHERE THEY ARE

This Golden Rule evolved from conversations with companies in the early days of the PSRT. Much of the initial debate centered around what is LCA and why should I care or even should I care. If I do, how do I respond? Some company members shared that they do not see any need to prepare LCA studies. Customers are not asking for it, nor are there any government regulations that affect them. However, there were a few company members who said that their management sees the potential value of LCA results to position them in the marketplace, even proactively, even if there are no immediate customers asking for it. Over time, and through several in-depth conversations with the company members, we developed four strategy levels (Figure 5.13).

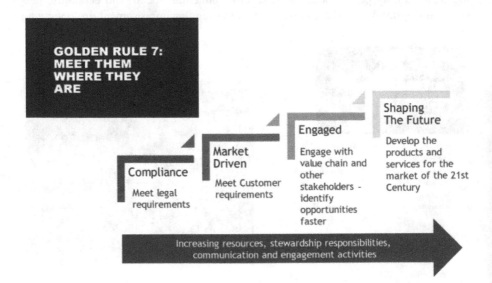

FIGURE 5.13   Meet them where they are.

- *Compliant level*: Meet legal requirements – *if I am forced to do so, I will.*
- *Market driven*: Meet customer requirements – *if a client asks for it, I will.*
- *Engaged/competitive*: Engage with value chain and other stakeholders – identify opportunities faster – *I see taking a proactive approach to the use of LCA results because we are in a competitive position.*
- *Shape the future*: Develop the products and services for the market of the 21st century – *I want to be seen as a leader in the space for both commercial and positive impact reasons.*

One aspect of this is to highlight those customers becoming a key partner is creating value.

Bring your customer in—It is natural in today's businesses to think in terms of value chains. However, for decades, companies have focused their environmental and social performance actions on operations that were directly under their control – e.g. their own faculties – and a lot of firms keep the lion's share of their attention there. For good reason, by the way, since your operations are where a lot of your costs and risks lie. However, **when a company shifts focus to their customer, amazing opportunities surface.** A customer focus requires looking at the full range of potential impacts along the entire value chain. Some steps in the chain are under direct control, others can only be influenced. Examples of upstream opportunities include building up better alliances, access to limited resources, stability of supply. Similar examples of downstream opportunities include customer satisfaction, loyalty, improved brand image, and reputation. We generally use the term "life cycle thinking" to describe this perspective. It is the concept that underlies the formal process of LCA, but it also is the thinking that drives sustainable materials management, resource efficiency, and circular economy as well. **Companies who intentionally bring customer focus into their sustainability strategy are better positioned to create and gain greater business value from their efforts.** A Harvard Business Review (HBR) article by Juan Richard Luciano (CEO of ADM) explained that their transformation over the last nine years was expanding its focus from commodities to consumers: this is an example of the importance and value of understanding your valued customers/consumers and structuring your business accordingly (Luciano 2023).

As a company moves from a complaint to a shape the future, there are increasing resources, stewardship responsibilities, communication, and engagement activities. Each of these strategic levels has a different degree of understanding and application of LCA results.

These four levels of strategy resonated well with the PSRT members. It helped them in their internal conversations about what LCA is and why I should care. Something a company may not have an explicit strategy within these terms. However, by asking and/or answering questions in relations about customers, government regulations, and customer actions, it informed the conversations to determine where they want to be with respect to LCA applications.

As firms move from a compliant strategy to a more sustainable strategy, different implications result. A *compliant* strategy, for example, is often viewed as a cost and often includes only strategic elements aimed at meeting the legal requirements as efficiently as possible. In a *market-driven* strategy, a firm has integrated pollution prevention and customer/consumer or reactive market considerations into the design of its products or processes, which results in cost savings and/or cost avoidance. On the other hand, *shaping the future* strategy may generate revenue by viewing the environment from a strategic perspective to identify new business opportunities and greener products. One advancement has been building alignment of the strategy levels and implementation of frameworks to various business values, that is, **growing revenue, enhancing brands, reducing costs**, and **mitigating risks**. These terms are not magic, nor locked in stone. Many companies have their own terminology and descriptions. For example, we worked with a company whose strategy was to be *first to be second*. Allowing another company to break ground. If successful, they wanted to be first to be second. The key is to have a strategy that is understood by all. It drives decisions and actions. Golden Rule 7 – meet them where they are.

GOLDEN RULE 8: FOCUS ON THE POSITIVE IMPACTS, NOT JUST THE DATA, INFORMATION, OR STUDY RESULTS

This Golden Rule illustrates the complicity of how product sustainability/LCA results are used, by whom and possible pathways those results are communicated and applied. As illustrated in Figure 5.14, positive impacts do not occur immediately after a study has been completed. It is a journey to translate results into impacts.

There is an interaction between the LCA practitioners and the professionals who are using the results to inform their decision-making. LCA results can be viewed as a first step. As we have shared, the users of the LCA results are engaged right from the start of the study. These users, depending upon the study's purpose, could be within marketing, customers, strategy, research, and development or stakeholder dialogue. During the goal and scope definition, these users share their intended purpose and how they might utilize the outcome of the study. They also share the business value

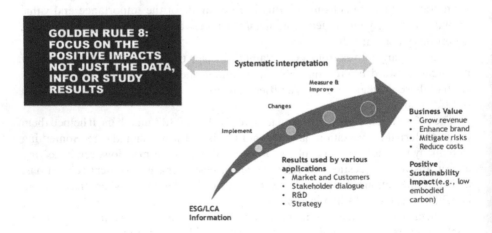

FIGURE 5.14 The purpose of the LCA results should create positive impacts.

that they would like to achieve. You will recall the bullets for each of the four major values – revenue, costs, risks, and brand. The brand manager is interested in communicating to an investor the sustainability of a new product. He/she is meeting the investor's expectations.

Once the LCA results are prepared, the LCA professional's role is not complete. We call their additional role a translator or interpreter role. They support the innovation professionals (or other business practices) to understand and apply the results. We shared, in Golden Rule 4, a few examples of how the LC professional can explain the LCT/LCA results in language more familiar with the innovation engineer.

Only after the LCA results inform decisions within the applications, and changes occur, can we realize the business values and positive sustainably impacts. We refer to this overall process as systematic interpretation, all within an accepted and proven governance process to utilize life-cycle information in innovation. This process may take time – e.g., in some cases a few years. That is why it is so important to ensure that the users of the ESG/LCA results are part of the G&SD right from the start; and the information providers maintain an ongoing linkage with the subsequent decision-making processes along the journey to value creation.

Remember, the outcome from sustainable innovation creates two very important and complementary values – business value and positive sustainable impact. If you are not achieving both, you need to look hard at what you are doing, why you are doing it, and how you are responding.

We asked one company we worked with when asked how they measured impact – he said they ask the businesses. Over the years, they built up an excellent collection of stories associated with moving from the study results to real business impact.

A couple of additional thoughts related to an engrained focus on the positive impacts, not just the data, information, or study results.

Understand the internal process and who are the key players/actors:

1. In order to scale and realize the potential business value, it is important to consider the steps, resources, and associated investments required to be successful.
2. As an example, you can consider an LCA report: If that report is put on a shelf and never used, its contribution to business value will be minimal; however, if the same report is used to inform R&D decisions, identifies key risks in the supply chain, is leveraged by the sales team in client engagements and is embedded in the marketing efforts the contributions can be substantial.
3. This graphic is simply a reminder that often there are additional steps between what we consider to be the "sustainability action" and the realization of business value and reduction of impacts.

Activities are important but only the first step to create actual positive impacts. Often, we consider providing data and information the result of our efforts. But it is often just the first step to achieve impact. Consider this framework (GEF 2009) for driving outcomes:

1. *Activity*: The indicator of our activity is a produced report – e.g., identification of hot spots for a product being redesigned.
2. *Outreach*: We reach out and train 20 product development professionals on the results and what they mean.
3. *Immediate impacts*: The next generation of the product for which the hot spots analysis was undertaken has reduced the impacts from each of the hot spots (e.g., use of energy, water and elimination of a material of concern).
4. *Longer-term impacts*: For example, the new redesigned product created an increase in 20% revenue over the previous product; and because of selling 500,000 units of this product, 1000 kWh as well as 1,000,000 gallons of water were saved.

It is important to understand what we want to achieve and how. The GEF report on "Impact Evaluation Review of Outcomes to Impacts Rotl" is a solid review and provide excellent insights on the path from activities to impacts (Source: "Impact Evaluation Review of Outcomes to Impacts Rotl" – GEF Evaluation Office with Conservation Development Centre – 2009).

BASF ingrained the use of sustainability approaches, e.g., sustainable steering solutions (SSS) and their eco-efficiency tool, into their innovation processes. Their SSS identified Accelerators that make a substantial sustainability contribution in the value chain. They set themselves an ambitious goal: "we will considerably increase the sales of Accelerator solutions to €22 billion by 2025 (2019: €15 billion)'. This target reflects our strong commitment to further drive innovation beyond today's state of the art." BASF has also calculated CO_2 footprints on all (45,000) sales products. Their approach recognizes the entire supply chain, not just manufacturing.

Sustainability and LC results are just the first step along a journey to realize sustainability and business positive impacts.

GOLDEN RULE 9: IT'S ALL ABOUT ACTIONS, CHANGING BEHAVIORS, AND BUSINESS PRACTICES

To achieve an organizations' vision or purpose, changes in actions and behaviors are necessary. These changes have to be embedded into the business practices, as we described in Golden Rule 6. Governance is the seat that makes the three legs (i.e., environment, social, and economic) of a stool. These actions to achieve the organization's goals/metrics are a formal part of the operational/business practices. If not, then when someone leaves/gets promoted, actions may vanish – go away. So, what do we mean by governance?

To achieve the full benefits of cost savings, improved brand, etc., a strategic approach for sustainability innovation efforts is necessary. The diagram (Figure 5.15) illustrates that without all the above elements in place, effective change is difficult to achieve. This is also true for successful innovation programs for sustainability performance. Leading companies have found that a truly comprehensive

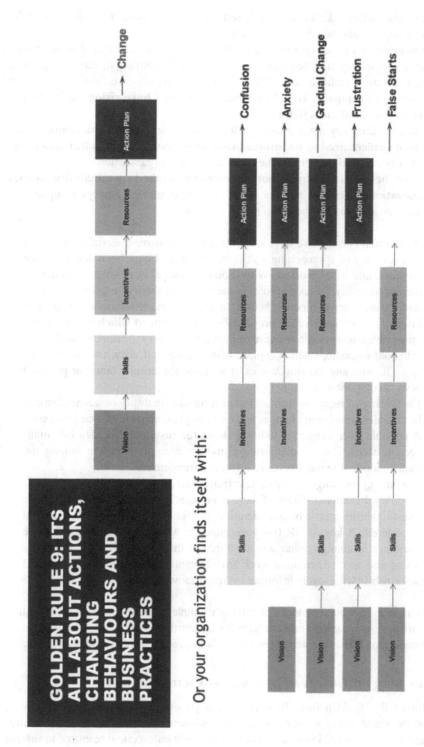

FIGURE 5.15   All must be in place to achieve change. (Adapted from Knoster, T. 1991. Presentation in TASH conference. Washington DC.)

strategy must further take a life-cycle-based approach to ensure that full scope of the product's impacts and benefits is considered.

When presented to senior managers, they all acknowledged that they have seen programs which they had been excited about end in confusion, anxiety, gradual change, frustration, or false starts. We are sure you have observed similar situations within your own company. **To achieve change we must have vision, skills, incentives, resources, and an action plan.**

To achieve this, key stakeholders including designers, product developers, and senior management need to be involved right at the beginning and throughout the project to develop and implement the specifics for any organization.

We have been asked, why do not all companies embed sustainability metrics into innovations? There is never a single or simple answer. Always complicated. Here are a few reasons and how you might respond.

- *Resistant to change*: We have seen when an alternative material is proposed or a new way of approaching a situation, we hear, I have been doing this for 30 years, and it has worked. Why should I change? A possible response by the sustainability/life-cycle professional is to show them examples illustrating how it works, or a competitor is using it and their market share has increased.
- *Lack of awareness*: A new proposal/idea is presented which is new or the innovative engineer is not aware of it. A possible way to address this barrier is to hold workshops which explain what is proposed, why, and how we can apply it, with any existing stories of successes internally (another product) or by a competitor.
- *Complexity of implementation*: Integrating sustainability or a new idea can be complex, involving changes across multiple departments and processes. A possible path forward is to break down the sustainability idea into manageable steps. Foster collaboration and open communication among the multifunctional teams to ensure a smooth transition.
- *Not part of our stage gate process*: Individuals' response is driven by their specific job responsibilities. If there is no mention or expectation to include a sustainability metric or consideration, it might just not get adopted or even considered. A key to GR 6 – governance is what drives change. The five attributes of change management shared in this Golden Rule 9, when followed and applied, create a work environment and expectations to embed sustainability/life-cycle information into innovation.

It is interesting, when one looks over these example barriers, the word collaboration often surfaces. As we shared earlier and will share further, working together to align on a common vision is the first step necessary to then agree on a path together to get there.

GOLDEN RULE 10: THERE IS A GROWING LIFE-CYCLE COMMUNITY

The final GR – Golden Rule 10 – There is a growing life-cycle community – all around the world – they are a resource – access them. Collectively the LC community and the users of LC information community are an excellent resource to inform and help you understand what, why, and how to accelerate the use of life-cycle information into decision-making to avoid burden shifting and at the same time to create both business values and help address societal needs. It is time to accelerate the use of the LC resource as a knowledge-based community to drive change and improvements.

We have come a long way since the early mass/energy balances, and the SETAC's LCA advisory group, who planned and carried out the early LCA workshops in the early 1990s. We have added credibility to LCA through the International Organization for Standardization (ISO) 14040 LCA standard series, built international and national understanding and capacity in LCA and LCT globally, helped to build capacity in governments and businesses globally (e.g., the UNEP/SETAC life-cycle initiative (now hosted by UNEP), created country/regional LCA support/networks, consulting companies, government and business professionals working in to advance life-cycle applications. LCA information is used within many sectors in their design guidance and planning. Universities have added LCA curriculum. LCT/LCA are being embedded as part of the engineering and business schools.

When I look back, it is amazing how far we have gone from less than 100 in 1990. If you consider LCT, LCA practitioners, and users of LC knowledge, my estimate is in the 10s of thousands. There are close to 700,000 product managers, who as they embed sustainability/LC metrics into innovation, and LCT/LCA is part of their processes, the number will expand even further.

One of the keys to this rapid growth and acceptance is the global nature of products and their life cycles, and the multi-impacts included as part of any LCT/LCA effort.

Having the ability to access and use that enormous amount of data to help innovation decisions is a no brainer – right thing to do. This brings up three opportunities for our resources going forward.

First, our technology software experts will continue to improve tools to provide data in the format and when it is most useful by the decision-makers in the best time frame within their processes. There is a lot of existing data. We must find the best ways to make it accessible.

Second, there will always be a need for more data. Innovation is continuing and in fact expanding. There will be new materials, new products, new markets, new regions, etc. Additional data will be needed. But tools and solutions will also be further enhanced to use existing and new data to inform decisions in the time required for those decisions.

Thirdly, as innovation continues to advance and as the number of engineers grow to provide the innovation know how's, the sustainability community who have been focused on providing the data, information, and/or software, there will be opportunities to evaluate and interpret that sustainability/LC data not as it related to the methodology, but as it related to informing the decision-making on its meaning to help them design more sustainable products. I see in the future greater opportunities for the sustainability/life-cycle professionals to become translators serving a niche between the data providers and the data/information users.

These are exciting times for the LCT/LCA and will now include the product managers and innovators to create sustainable products.

After 30 years of the modern era of LCA, we know what LCA is, its capability has been improved and enhanced through the collaborative efforts within ISO and other standard setting efforts (EPD, PCRs), as well as the numerous NGO, consulting, governments, businesses, universities, and other organization who continually enhance and apply LCT and LCA. Hopefully these Golden Rules will be helpful in those efforts.

WHERE DO WE GO FROM HERE – SUSTAINABLE INNOVATION IS THE NEXT FRONTIER FOR GREENER PRODUCTS?

Where do we go from here? To me, the answer is "**Sustainable innovation is the next frontier for greener products**." Sustainability associated with products is becoming a powerful driver to change companies' strategies, brand/business practices, and financial performance. Products will be designed with sustainability attributes as a core part of the innovation process. Sustainability knowledge will be at a similar par as safety, performance, costs, reliability, and other key attributes within the innovation processes. A successful process will be the ability of the innovation team to take action to scale the use of life-cycle information, delivered on time, to influence decision-making to generate more sustainable products. Foundational to our use of the term sustainable innovation is that it includes both environmental and social attributes.

With this perspective, there are a couple of enablers which will enhance this vision to be realized.

- Rapid increase in government (EU and US government) expectations for more sustainable products.
- Sustainability professionals will be an integral part of the innovation teams, so that sustainable products will be co-created with a significant contribution from the sustainability professionals.
- LCA practitioners, designers, business managers work together to gather and share knowledge, tools, and processes.
- LCT is applied all the time.
- LCA is the core tool to provide the knowledge to inform design decisions to create more sustainable products.
- LCT/LCA contribute to identifying hot spots/opportunities to drive more sustainable products/businesses.
- Social responsibility for your support chain is an integral part of the Innovation process.
- Collaboration with stakeholders (e.g., recycling, recover organizations) to manage hot spots for the product that may occur after launch.

This enhanced knowledge is suited to the multi-stages along NPD processes or other core business functions like procurement, marketing, policy development.

- Resulting in sustainable information is embedded into the innovation process and other relevant policy and business practices.

Only by collaborating on the next generation of knowledge, tools, and their applications will we scale the use of life-cycle information, delivered on time, to influence decision-making to generate more sustainable products and businesses. Co-creation is not new – in business, engineers have been co-creating as part of their normal operations. Embedding sustainability into innovation is critical to advance additional co-creation imperatives. Collectively, then, sustainability including life-cycle knowledge are available and are part of the innovation process. The outcomes are sustainably innovated products with reduced environmental impact, minimized social impacts, enhanced positive sustainability impacts, and positive business values.

You may ask, with the many governments, investment community, leading companies, and social expectation for sustainable actions by companies, why do we need to provide additional priority to sustainable innovation? Several overarching developments and observations have surfaced that have created a very high demand to scale and accelerate embedding sustainability into innovation.

In our previous chapter (Edition # 2), we shared the example of a keynote speaker who did an outstanding job in illustrating and demonstrating that businesses that perform well from a sustainability perspective also perform well from a business perspective. This was not a new message. It is conveyed almost daily on social media. Then, the speaker was asked, if the case is so strong for companies to create business value from operating sustainably, why isn't everyone doing it? Clearly a million-dollar question. Last edition, we talked about speaking the language of the user of the sustainability information (our Golden Rule 4) and approaching the decision-maker recognizing where they are in their understanding of sustainability (our Golden Rule 7), and then proceed from there.

Those two points are still valid. Based upon the experiences over the last five years, others surfaced which are also critical to be understood and acted on for companies to move forward.

- *Sustainability issues are not a single impact*—That should be obvious. But we have too many single impact strategies, commitments (e.g., climate change, biodiversity). Sustainable innovation means that innovation addresses multiple environmental and social impacts at various stages along the product's life cycle. Translating those multi-impacts into actionable attributes is not an easy process, given the decision-making time constraint, engineers are faced within their innovation process. Examples of how companies, organizations are addressing these multiple impacts and the language of the decision-makers have been outlined in Golden Rules, the case studies, and from the multi-organization project to bring LC information to the beginning and throughout innovation. Clearly, more information and examples of success stories are needed across many sectors and global regions.
- **Impacts that are identified via HSA often occur in the future**, where any actions necessary to solve and address an impact are not in the 100% control of the innovation team. For example, end-of-life management of solar panels, windmills, and automotive batteries occur 10–25 years in the future. While these longer-term impacts are recognized and acknowledged,

it requires the shared responsibility of others who can act on them at the right time. This does not take the responsibility away from the innovation team, but it does add complexity. As we discussed earlier, to fully meet the direction clearly defined by the Brundtland report, we must identify and engage the stakeholders who can control and/or influence those hot spots over the entire life span of the product, even if the hot spot is 25 years in the future. The product managers also have a role in addressing what can be done within the design of the product and work with the other multi-stakeholder organization who will be able to deal with the end-of-life impacts of the product.

- *There is a rapidly increased concern about the social issues associated with suppliers* —Sustainability is more than the environment. Social issues include worker conditions and child labor to name a few. The LC community has been working to advance social impacts to be an integral part of LCT/LCA approaches (UNEP LCI reports, 2022). The sustainability community has been driving social improvement within the supply chain for a while. Social performance of suppliers is becoming a key issue in the innovation process.
- **Customers and consumers are not only showing interest in "greener products" but are acting on that interest**. Two examples:
  1. Most Consumers Want Sustainable Products and Packaging (https://www.businessnewsdaily.com/15087-consumers-want-sustainable-products.html). Two key takeaways from a recent article
     a. Many customers are **willing to pay more for sustainable products** with high-quality, environmentally friendly packaging.
     b. Consumers are more **attracted to brands with sustainable practices and products** – and those businesses turn a profit.
  2. Products making **ESG-related claims averaged 28 percent cumulative growth over the past five-year period**, versus 20 percent for products that made no such claims" (McKinsey and Company and NielsenIQ 2023) (https://nielseniq.com/wp-content/uploads/sites/4/2023/02/Consumers-care-about-sustainability%E2%80%94and-back-it-up-with-their-wallets-FINAL.pdf).
- *Governments are imposing their authority to develop additional reporting, procurement, and sustainable products and services directives*. For example, the US government proposed to establish requirements for prioritizing sustainable products and services. Specifically, when procuring sustainable products and services, agencies shall ensure compliance with all applicable statutory purchasing programs and prioritize multi-attribute sustainable products and services, i.e., that meet the requirements of both a statutory purchasing program and a required EPA purchasing program. Similarly, making the EU a more resource-efficient, climate neutral and pollution free circular economy calls for more sustainable products (https://commission.europa.eu/energy-climate-change-environment/standards-tools-and-labels/products-labelling-rules-and-requirements/sustainable-products/about-sustainable-products_en#overview).

- *Companies are exploring where the sustainability role is placed within their organization*. In the opinion of the author, innovation and procurement will become a priority role for actions to embed sustainability into their business practices. This does not diminish or reduce the importance of enterprise sustainability efforts and reporting. It does create an increased effort on sustainable products and services. Enterprise sustainability is essential but not sufficient without product sustainability.
- Finally, with the rapidly changing social media world, the **advancements in digital technologies and AI**, the work forces (both current and future) will recognize that with these changing technologies, their jobs and work environment are changing. What they learned 5–10 years ago is rapidly changing. From my perspective, it provides an excellent opportunity for both the providers of sustainable and life-cycle information, AND the users of that information, to expand their evaluation and interpretation skills and insights. AI will be able to provide the information. A professional's value will be able to translate and help the users to understand and be better prepared to incorporate that information into their decision-making process. It is certainly a scarier time for many professionals, but it is also an exciting time to accelerate the ability of all the information and efforts we have spent over the last several decades to be better understood and used in innovation. With social media and the ability to share facts globally, the younger generations will be able to transition to new, and from my perspective, an exciting period to be able to translate the sustainability/life-cycle information to knowledge to be used by decision-makers.

In the September/October 2023 issue of the *Harvard Business Review*, there were two papers that provided solid recommendations, consistent with the messages in this chapter (Tamayo et al. 2023; Martinez 2023).

Tamayo et al. (2023) defined five paradigms for leaders and employees to **reskill in the age of AI**. They are:

- Reskilling is a **strategic imperative** – built into the core strategy for growth and development.
- Reskilling is the **responsibility** of every leader and manager – not just Human Resources.
- Reskilling is a **Change Management Initiative**. As I shared with my Golden Rule 9, sustainable innovation is also about change management.
- Employees want to reskill when it makes sense – i.e., LCT to be used all the time, LCA when it makes sense.
- Reskilling takes a village – partnering with – e.g., industry organizations, non-profits, local colleges, and training providers – e.g., SD goal 17 – partnership.

While that article did not address sustainable innovation, these five paradigms to reskill for AI, apply to reskilling for Sustainable Innovation.

Martinez (2023) completed an interview of Professor Timothy DeStefano and colleagues on how people were trusting or not trusting AI when they could not see how it worked. An answer by Professor DeStefano really resonated with me. Question: "if you were introducing an AI-based recommendation tool to your employees, would you make it uninterpretable." Answer by Dr DeStefano, "Before I roll out a new tool, I'd have the data science team or whoever was building it meet with people who would be using it and include some of them in the development process." It is interesting, the LCA communities have been incorporating the users of the LCA results right from the start – i.e., goal and scope definition (first stage of an LCA) and throughout the LCA process. For our focus in embedding sustainability into innovation – the implications are that the users understand the process, where the data come from and how the results can be incorporated into their innovation process.

The Professor's final statement is one that clearly shows what the sustainability and life-cycle community and the user community needs to work on. The statement was "… the system had been tested and everyone agreed that it is working, I'd want to limit my employees' ability to overrule the algorithm." While LCA or another sustainability tool is nowhere near an AI tool, the idea of having the users accept the credibility and value of the outcomes is a challenge for us and should be a focus to the AI community to better explain what is behind the AI tool, in the language of the audience (i.e., Golden Rule 4).

Putting these observations together, the value chain becomes a priority, which make up a significant component within a product life cycle. It has been shown (World Wildlife fund – WWF) that **up to 80% of the product life-cycle costs/ impacts are determined in the product design stage** – although the actual costs occur downstream. When one reviews the (a) typical time spent by management with the (b) opportunities for value creation (e.g., through environmental and social responsibility) with time actual spent by management, we see a few things:

- It is often the case that 80% of the current management efforts covers only 20% of the opportunities for value creation – i.e., manufacturing.
- While 80% of the value creation opportunities come from:
  - upstream opportunities (e.g., building better alliances, access to limited resources, stability of supply…) and
  - downstream opportunities (e.g., customer satisfaction and loyalty, improved brand image, …).

Collectively, this is why there is an accelerating focus and priority on "Products" including their entire value chain. What does that mean to a company?

- Product managers will become leaders in the transformation to sustainable products. In one survey, there were around 700,000 product managers.
- Product managers are responsible for guiding a product or service throughout the product life cycle. The product life cycle covers the discovery and development of the product, its introduction to the market, and its subsequent growth, maturity, and decline.

- The product manager is the go-to person for anything relating to the product. They collaborate closely with many different stakeholders across the business – including the leadership team, designers, developers, sales and marketing, and customer care teams. Ultimately, anyone who is involved with the product or service in some way will be in contact with the product manager.
- The product manager provides a perfect role to embed social and environmental factors that can address risks or opportunities.
- Product systems and value chains become a key priority for change and improvements. Companies make the money by selling products – great opportunity.
- Innovation, product development, commercialization is where the breakthrough innovation and improvements for more sustainable products will occur.
- Life-cycle knowledge will be embedded into business/government processes – e.g., innovation – used by product managers. The LCA community has a role to help translate their results to the language of the user (Golden Rules 4 and 8).

CONCLUSION

We close the chapter with a call for action. Applying the new and additional insights shared related to Golden Rules and learnings from my career and the collaborative project to bring life-cycle information to the beginning of innovation created clarity on a vision for actions to scale the development of more sustainable products. These include deeply embedding sustainability metrics into innovation and procurement. Sustainability and life-cycle professional's roles will evolve to also include translators, interpreters, and knowledge sharing, and illustrating how a sustainable agenda creates business values. Moreover, people skills complement the technology/tools being developed and applied to become important. One of the key walks away messages is the need to move beyond what is a sustainable product and why, to the who and how. These insights outlined hopefully help product managers, and others with key roles within companies, to take positive actions NOW to develop more sustainable products.

Now is the time for sustainability including the life cycle, and the user communities to come together to co-create the next generation of tools and processes, which deliver sustainability/life-cycle results in time, and in a way that is understood, and usable by the user communities. With the excitement, enthusiasm, and commitment of the current and upcoming generations, this will happen. Be part of the transformation. Do it now!

REFERENCES

AHAM 7001-2012/CSA SP. E-7001-12/UL. 7001. 2012. The Sustainability Standard for Household Refrigeration Appliances. https://webstore.ansi.org/standards/csa/csaspe70012012. Accessed (May 1, 2024).

Anthesis. Life Cycle Assessments: Converting Results into Business Value. https://www.anthesisgroup.com/life-cycle-assessments-converting-results-into-business-value/ (Accessed December 10, 2023).

BASF. 2017. Sustainable Solution Steering. file:///C:/Users/aliannuz/Downloads/BASF_ Sustainable_Solution_Steering_Manual%20(2).pdf (Accessed May 1, 2024).

EPEAT Criteria in NSF 457 – Sustainability Leadership Standard for Photovoltaic Modules and Photovoltaic Inverters. https://globalelectronicscouncil.org/wp-content/uploads/ NSF-457-2019-1.pdf. (Accessed May 1, 2024).

Fava, J. 2021. Getting Sustainability Right: 10 Golden Rules for Applying Life Cycle Information to Create a Sustainable Business. https://fslci.org/wp-content/uploads/ youzify/groups/12/2021/03/Golden-rules-for-a-sus-bus-Fava-NZ-11-March-2021-final. pdf. (Accessed April 29, 2024).

Fava, James. 2012, 2018. Framework for Developing Greener Products. In Al Iannuzzi, *Greener Products: The Making and Marketing of Sustainable Brands.* 1st Edition. CRC Press, Boca Raton, FL.

Footprinter. https://footprinter.com/blog/item/70c380658961fe453bb7cb34170c0cc3 (Accessed May 2, 2024).

GE. 2017. *A Decade of Ecomagination.* http://dsg.files.app.content.prod.s3.amazonaws. com/gesustainability/wp-content/uploads/2015/01/Ecomagination-Timeline-31.png (Accessed January 25, 2017).

Global Environment Facility (GEF). 2009. *The ROtI Handbook: Towards Enhancing the Impacts of Environmental Projects – Methodological Paper # 2.* https://www.gefieo. org/sites/default/files/documents/ops4-m02-roti.pdf. (Accessed May 2).

Iannuzzi, Al. 2012, 2018. *Greener Products: The Making and Marketing of Sustainable Brands.* 1st & 2nd Editions. CRC Press, Boca Raton, FL.

Luciano, Juan Ricardo. 2023. How we did it. *Harvard Business Review.* September/October, pp. 32–35.

Martinez, J. 2023. People may be more trusting of AI when they can't see how it works. *Harvard Business Review.* September/October 2023, pp. 30–31.

McKinsey and Company and NielsenIQ. 2023. Consumers care about sustainability—and back it up with their wallets. https://nielseniq.com/wp-content/uploads/sites/4/2023/02/ Consumers-care-about-sustainability%E2%80%94and-back-it-up-with-their-wallets-FINAL.pdf.

NielsenIQ. 2023 https://nielseniq.com/global/en/landing-page/tl-the-changing-story-of-sustainability/ (Accessed May 2, 2024).

Perry, Stott. 1993. Responsibility Products. *Smallwood.*

Si, et al. 2023. New approach to strategic innovation. *Harvard Business Review,* September/ October, pp. 120–129.

SSCA. 2016. State of the Alliance 2016. Presentation, SSCA Supply Chain Sustainability Conference. New Orleans, LA, September 29, 2016.

SSCA 2018. 2018 goal progress dashboard. (From SSCA, state of the Alliance 2016, Presentation, *SSCA Supply Chain Sustainability Conference,* New Orleans, LA, September 29, 2016. With permission.).

SSCA. 2023a. Sustainable Supply Chain Alliance Sustainability Framework. https://www. tspproject.org/. (Accessed May 2, 2024).

SSCA. 2023b. "About Us" Alliance Website. https://community.thessca.org/home/ssca-about-us. (Accessed May 2, 2024).

Tamayo et al. 2023. Reskilling in the age of AI. *Harvard Business Review.* September/ October, pp. 56–66.

UL. 2012. htttp://www.ul.com/global/documents/offerings/businesses/environment/press/ ULE_whirlpoolcertifies.pdf (June 14, 2012) (Accessed January 25, 2017).

UNEP. 2014. *Hotspots Analysis: An Overarching Methodological Framework and Guidance for Product and Sector Level Application,* Fava, J. and Barthel, M., co-chairs, May 2017, Paris, France. http://www.lifecycleinitiative.org/new-hotspots-analysis-methodological-framework-and-guidance/.

UNEP. 2017. UNEP/SETAC Life Cycle Initiative—Flagship Project 3a (Phase 2). Hotspots Analysis: An Overarching Methodological Framework. Fava, J. and M. Barthel, co-chairs. https://www.lifecycleinitiative.org/wp-content/uploads/2014/10/Flagship3a-Hotspots-Mapping.pdf. (Accessed May 2, 2024).

UNEP LCI reports. 2022. Guidelines for social life cycle assessment of products and organizations. https://www.lifecycleinitiative.org/library/?filter_data-and-methods=social-lca report. https://www.lifecycleinitiative.org/library/pilot-projects-on-guidelines-for-social-life-cycle-assessment-of-products-and-organizations-2022/ (Accessed December 10, 2023).

Walsof Pty Ltd. 2003. *AcdB™ Questions and Outputs*. 2nd Road Pty Ltd.

6 Biodiversity and Natural Capital

Libby Bernick

INTRODUCTION

The consumer and capitalist worlds are on a collision course with the natural world. Consumer demand for products is set to grow in the next 20 years with the increase in middle-class consumers, especially those in emerging markets, who will require food, clothing, housing, and transport at levels never seen before. Companies making these products rely on nature, whether it's the water needed to grow cotton for our clothes or the trees that provide wood for housing materials. Alongside the continued growth in consumption, the world is grappling with a changing climate and its knock-on effects, like increased droughts or wildfires. Biodiversity, the variety of all living things on our planet, is rapidly declining mainly due to human activities, such as pollution, climate change, and clearing land for food production or development. The authoritative review by Professor Dasgupta on the *Economics of Biodiversity* (Dasgupta Review 2021) spells out that demands on nature far exceed its capacity to supply them, putting the planet under huge pressure and society at "extreme risk."

Simply put, nature is an important part of the global economic engine, the brands fueling it, and the investors that provide equity or debt financing to those brands. Companies and their investors are increasingly under a spotlight to show how their actions and investments affect the planet. Successful business models will be those that understand the value of natural resources—commonly referred to as natural capital—and where there are business opportunities to create greener products and services that solve environmental challenges, like climate change and biodiversity loss.

BUSINESS AND INVESTMENT IMPLICATIONS

To illustrate the magnitude of these risks and opportunities, Impact Cubed analyzed global equity markets using its biodiversity data, which in Figure 6.1 shows that over 35% of global companies representing over $28 trillion in market capitalization are exposed to biodiversity risk. This significant exposure extends to companies in both developed and emerging markets, with over $19 trillion at risk in developed markets and nearly $9 trillion in emerging markets.

In 2020, the Organisation for Economic Co-operation and Development (OECD) estimated that $44T of economic value generation—over half of global GDP—moderately or highly depends on nature (Nature Risk Rising 2020).

DOI: 10.1201/9781003441939-8

FIGURE 6.1   Biodiversity market capitalization exposure. (Source: Impact Cubed.)

What's more, out of that $28 trillion market cap exposure, over $25 trillion is exposure to products and services that are contributing negatively to the drivers of biodiversity. This highlights the potential financial business risk associated with continued biodiversity loss.

The 2013 landmark Trucost study, Natural Capital at Risk—the Top 100 Externalities of Business, researched primary production industries (agriculture, forestry, fisheries, mining, oil and gas exploration, utilities; primary processing industries: cement, steel, pulp and paper, and petrochemicals) and tallied the cost to the economy around $7.3 trillion a year—about 10% of the global economy at the time (TEEB 2013). Although the total costs were strikingly high, the starker revelation was that the natural capital cost is higher than the total revenue of each sector. In other words, **not one of the top 100 businesses would be profitable if it had to pay the environmental costs associated with production!**

The economic costs are from a range of impacts, including things like:

- carbon pollution,
- health costs and societal damages from air pollution (for instance, asthma or crop damage),
- water pollution from overuse of fertilizer which creates a cost for businesses or communities who have to pay to treat the water so they can safely use it, or
- water exploitation in areas where it is scarce.

The analysis was the first to quantify and value "**hot spots**"—areas of greatest environmental impact for the economy broken out by region and commodity—showing that the highest environmental costs are associated with energy generation and food production.

More recently, in 2020, the OECD estimated that $44T of economic value generation—over half of global GDP—moderately or highly depends on nature (Nature Risk Rising 2020). An obvious example is the **forest products industry which is 5% of the total US manufacturing GDP, makes over $300 billion in products annually**, and provides more than 900,000 jobs (American Forest and Paper Association 2016). Less obvious are sectors like technology, whose data centers rely on massive amounts of cooling water to provide an uninterrupted stream of digital content. Or for example, the clean beach that an airline might depend on to book seats to resort destinations.

While many natural capital costs occur in emerging markets, the resulting products are consumed by supply chains in developed economies, making it a global challenge for a globalized world. The Nature Risk Rising report categorizes how different regions depend on nature, as seen in Figure 6.2. Fast-growing economies are highly sustained by nature's assets and are also at risk of severe environmental damage. For example, a third of India's and Indonesia's GDP is heavily reliant on nature, along with the African continent at 23%. Economies with the highest GDPs like China ($2.7 trillion), EU ($2.4 trillion), and the United States ($2.1 trillion) have significant environmental impact (Nature Risk Rising 2020).

WHAT IS DRIVING BUSINESS INTEREST IN BIODIVERSITY AND NATURAL CAPITAL?

The variation in size and type of impacts across regions and commodities suggests companies have a very real opportunity to differentiate their brands, business models, and manufacturing processes by considering natural capital. Supply chains, sourcing decisions, operating locations, and product designs can all be informed and optimized with better understanding and measurement of natural capital dependencies.

The main reason companies do not innovate for nature-positive solutions or address biodiversity in their decision-making is that natural resources are largely common goods, underpriced (or unpriced) by today's economic system. **The costs of pollution and exploitation of natural resources are, for the most part, not on the balance sheet or profit and loss statement**, discouraging companies and investors who need immediate market returns from investing in innovations that would prioritize greener products and supply chains. Ford example, hydraulic fracturing to extract natural gas uses copious amounts of groundwater, which in high-production regions of the United States like Texas can be withdrawn, used, and redisposed in aquifers at no cost—leading to depleted and contaminated groundwater supplies for other users.

The situation is akin to having a group of high-use commodities that drive enormous economic value, but where there are no price-setting participants because there is no market. Then consider the potential for mispricing and dramatic price volatility of these resources and the implications for business performance. In the most extreme case, the resources disappear before the price can move because there is no price signal to curtail demand. That is the precise situation for the vast majority

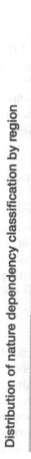

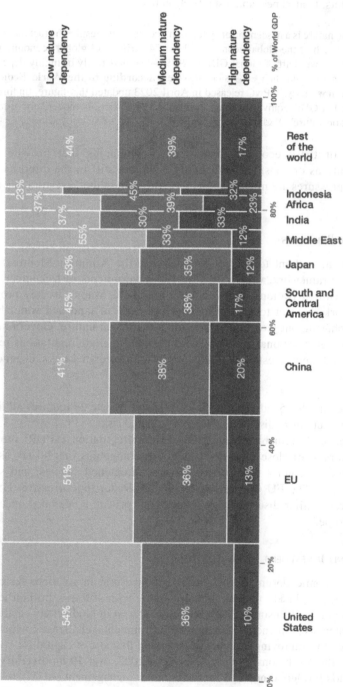

Distribution of nature dependency classification by region

Low nature dependency

Medium nature dependency

High nature dependency

% of World GDP

Rest of the world — 44% / 39% / 17%

Indonesia — 23% / 45% / 32%
Africa — 37% / 39% / 23%
India — 37% / 30% / 33%
Middle East — 55% / 33% / 12%
Japan — 53% / 35% / 12%
South and Central America — 45% / 38% / 17%
China — 41% / 38% / 20%
EU — 51% / 36% / 13%
United States — 54% / 36% / 10%

Source: PwC

FIGURE 6.2   Distribution of nature dependency.

of the world's essential environmental resources, from clean air and fresh water to natural habitats, landscapes, and a stable climate.

> "Protecting nature is a nascent priority for many companies, despite the fact that nature underpins much of the global economy. About $44 trillion of global economic value generation — over half of global GDP in 2019 — is moderately or highly dependent on natural assets and their ecosystem services, **according to the World Economic Forum**. A **new analysis** PwC released in April 2023 updated that figure, finding that 55% of global GDP, equivalent to an estimated $58 trillion, is moderately or highly dependent on nature" (S&P Global 2023).

Because of these economic headwinds, companies that provide nature-positive solutions or consider biodiversity risks are still in the minority. At the same time, pressures are mounting on companies and their investors in a number of ways.

REGULATORY PRESSURES

A landmark agreement to protect biodiversity, the Kunming-Montreal Global Biodiversity Framework, was ratified by 190 governments in December 2022 at the COP15 United Nations biodiversity conference. Among its key provisions, the framework sets out to protect and restore 30% of land and water by 2030 while also phasing out government subsidies that harm nature. Governments are expected to create national action plans that could include regulations requiring companies and their investors to publish their nature-related risks, dependencies, and impacts.

- France Article 65, adopted in May 2021, and Article 29 requires all financial institutions to disclose biodiversity-related risks.
- European Sustainable Finance Disclosure Regulation (SFDR) requires investors to disclose their investments in companies that do not assess, monitor, or control biodiversity drivers, threatened species, and sensitive areas. The EU Corporate Sustainable Reporting Directive (CSRD) requires similar disclosures by public and private companies operating in Europe.

BUSINESS AND INVESTMENT OPPORTUNITIES

A World Economic Forum Report finds that investing in solutions beneficial to nature can help yield $10 trillion annually, along with 400 new million jobs (The Future of Nature and Business 2020). Investor interest in biodiversity capitalizes on investors' interests in financing solutions to environmental challenges, which creates opportunities for improving liquidity or lowering the cost of capital for companies that provide these solutions. For example, as of 2022, over **10 biodiversity-related mutual funds totaled about $1B in assets** under management (AUM), reflecting a wide range of listed companies that make up their holdings as shown in Figure 6.3 (Stewart 2023).

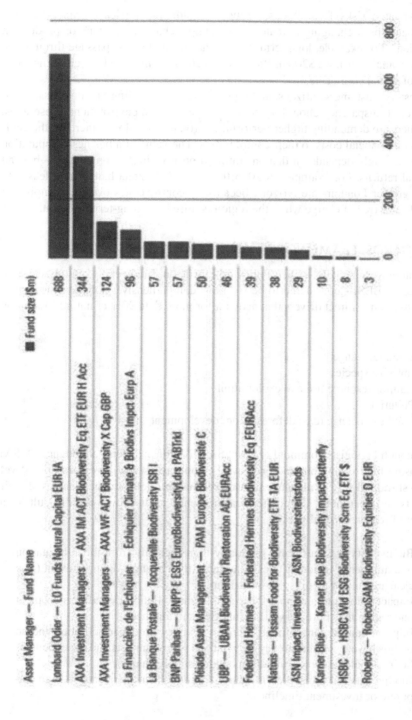

FIGURE 6.3 Biodiversity mutual funds.

BUSINESS AND REPUTATIONAL RISKS

Nature-related risks from floods, droughts, wildfires, hurricanes, and the like, exacerbated by a changing climate, create supply-chain disruptions or production limitations. For example, long-term drought has limited ships' passage through the Panama Canal, causing a $200 million drop in 2023 earnings and affecting the move of 6% of global commerce.

Investors, customers, citizens, and regulators are demanding that businesses provide more transparency about how they manage or harm environmental resources. Customers are demanding higher-performing products that do not increase the environmental or social costs to their communities. The value of a business's reputation also increasingly depends on demonstrating a positive impact on society—beyond financial returns. For example, in 2023 the state of Arizona halted land leases to alfalfa grower Fondomonte Arizona because of controversies over its pumping of groundwater free of charge while the region is mired in a long-term drought.

STANDARDS, FRAMEWORKS, AND COALITIONS

The Intergovernmental Science-Policy Platform on Biodiversity and Ecosystem Services (IPBES)—considered to be the most comprehensive global assessment to date—found five direct drivers that account for more than 90% of nature loss in the past 50 years:

- Climate change
- Invasive species
- Natural resource use and exploitation
- Pollution
- Land use change (e.g., deforestation, development)

Although biologists commonly use metrics like mean species abundance (MSA) to measure biodiversity on land or in water, **companies and investors do not yet have a standard way to measure nature-related impacts** or dependencies of their products and operations. Corporate measurement and disclosure are difficult for a number of reasons, including:

- Biodiversity loss can occur gradually then suddenly from the culmination of a number of incidents over time rather than a single defining event (like clearing land or an oil spill, for example).
- Impact is local to a region, for example, air pollution that affects crops in one region may have very different effects in another.
- Impacts can be concentrated more in primary producer activities and so brands have less transparency on what and where impacts and dependencies lie in globally complex and often opaque supply chains.
- Biodiversity risks may present themselves on longer time horizons than corporate or investment timelines.

TASK FORCE ON NATURE-RELATED FINANCIAL DISCLOSURES

The Task Force on Nature-Related Financial Disclosures (TNFD) has emerged as a leading **voluntary risk management and disclosure framework**, highlighting eight non-financial sectors more likely to experience financial impacts related to biodiversity loss due to both their strong dependencies on nature and the effects they have on it. About 35% of the stock market MSCI World Index consists of companies in the TNFD priority sectors, including food and beverage, extractives, transportation, and others.

SCIENCE-BASED TARGETS NETWORK

The Science-Based Targets Network (SBTN) has launched the first science-based targets for nature, aiming to set the global standard for measurable corporate action to protect nature. The network includes an initial group of 17 global companies (AB InBev, Alpro (part of Danone), Bel, Carrefour, Corbion, GSK, H&M Group, Hindustan Zinc Limited, Holcim Group, Kering and L'Occitane Group, LVMH, Nestlé, Neste Corporation, Suntory Holdings Limited, Tesco and UPM) providing leadership and practical tactics for action.

Natural Capital Accounting

Natural capital accounting emerged in the 1970s as a tool for conceptualizing how society impacts and depends on nature by placing a monetary value on the activity. Natural capital accounting classifies nature into "stocks" (such as land), which can be renewed (forests) or are non-renewable (fossil fuels). Nature-related "services" flow from these stocks.

Natural capital is an economic metaphor for the total stock of physical (non-living) and biological (living) stocks, and their capacity to provide services to people. Higher biodiversity emerges from healthier natural capital, in practice making it a subset of natural capital. Some economic benefits provided by natural capital are valued by the market with a price, for instance, commodities like wood. Other goods and services that nature provides are equally important to the economy yet not fully priced or valued by markets; for example, clean water that a business needs to run its production processes or the clean beach that an airline might depend on to book seats to resort destinations. The term **natural capital** cost is used here to **describe the non-market value of the environmental resources that businesses depend on to grow revenue**.

There are many approaches to measure and value natural capital as shown in the following table, which illustrates a range of organizations applying natural capital accounting at different scales and for different purposes. At the same time, public and private sectors are working to harmonize and align approaches for natural capital accounting, which will be required if these tools will be scaled across industry sectors and investment practices.

Examples of Natural Capital Valuation

Scale or Type of Application	Example Initiatives
Sovereign	Canada, Australia, the United Kingdom, and the United States are now adopting accounting systems for adding natural capital to economic statistics to track the stock and flows in line with UN System of Environmental Economic Accounting (SEEA). Wealth Accounting and Valuation of Ecosystem Services (WAVES 2016), a global partnership led by the World Bank
Financial Lending	In 2015, TD Bank valued the natural capital impact related to reduction of greenhouse gas (GHG) emissions and airborne pollutants of TD Bank's $200 million green bond (TD Bank 2015)
Commodity	American Chemistry Council in 2016 valued the costs and benefits of plastic consumer products and packaging (American Chemistry Council 2016)
Corporate Enterprise	Kering (the French luxury goods holding company which owns Alexander McQueen, Balenciaga, Gucci, Puma, Volcom, and other sport and lifestyle brands) developed the concept for an Environmental Profit and Loss framework in 2011 and has since applied it across the operations and supply chain of its enterprise (Puma 2024
Product	Dell in 2015 valued the net benefits of its closed-loop plastic recovery system (Dell 2015)

Why Value Natural Capital?

There are a range of reasons why organizations use monetary valuation tools to assess their impacts and dependencies on nature, including:

- *Quantify environmental impacts in business terms*: Putting a monetary value on impacts translates environmental issues into business terms and allows businesses to consider their environmental costs alongside financial costs.
- *Provide context for investors, customers, and consumers*: Communicating about the benefits and impacts of sustainable brands in monetary terms helps engage customers and consumers who are not environmental experts. For many people, nature can be a bit abstract. Seeing the true cost of a product alongside the financial cost shows that these environmental impacts are real, they are being paid for by society, and they drag down our economies.
- *Evaluate trade-offs*: There are different types of environmental impacts (e.g., air pollution or water pollution) and each is expressed in a different unit, from CO_2 equivalents to gallons of water used. For example, if greener brand A has higher greenhouse gas emissions but lower waste compared to greener brand B, which is better? **Natural capital valuation places a monetary value on each impact**, so they are all expressed in financial values. In this way, impacts can be added and compared, and trade-offs between different design strategies become clear.

- *Science-based, context-based insights*: Natural capital valuation is also a way to communicate about more sustainable brands in an objective way. Because natural capital valuation accounts for the availability of the resource, business can understand the context for its resource use. For example, a company that uses water from a stressed aquifer has a much greater impact than if water is used from a plentiful region. In addition to providing science-based environmental context, environmental costs are presented in a business context that allows companies to understand how material costs and risks impact future profitability.
- *A proxy for risk*: Externality costs are an excellent proxy for business risk. Natural capital costs affect the future profitability of businesses if they are internalized as a business cost because of increased regulation (e.g., a tax on carbon emissions or reduced water allocations). These internalized costs must then be absorbed by reducing the profits paid to shareholders or (more likely) passed on to consumers. These natural capital costs are especially important for industry sectors with small profit margins.

Natural Capital Valuation Frameworks

One of the first frameworks developed was the groundbreaking **Environmental Profit and Loss framework created by Puma in 2011** (see Figure 6.4), intended to mirror a conventional Profit and Loss statement by quantifying environmental costs in monetary terms. For Puma, environmental costs totaled €145 million and were largely concentrated in cattle-rearing and cotton-farming activities in the upper tiers of its supply chain (Puma 2024). This type of framework is especially well suited to understanding corporate enterprise-wide risks because it aggregates natural capital costs across all operations and supply chains and can be broken out by environmental impact, business unit, region, or commodity. Since established, **the framework revealed that more than 50% of environmental impact is found in the material and raw-material production stage**. To combat such problem, Puma set 90% of their 2020 environmental targets to use more sustainable raw materials (Puma 2024).

EUR million	Water use	GHGs	Land use	Other air pollution	Waste	Total	% of total
	33%	33%	25%	7%	2%	100%	
Total	47	47	37	11	3	145	100
PUMA operations	<1	7	<1	1	<1	8	6
Tier 1	1	9	<1	1	2	13	9
Tier 2	4	7	<1	2	1	14	9
Tier 3	17	7	<1	3	<1	27	19
Tier 4	25	17	37	4	<1	83	57

FIGURE 6.4   Puma Environmental Profit and Loss (EP&L) framework.

Regardless of the framework that is applied to organize or communicate the insights, the natural capital valuation methodology generally follows three steps:

- Quantify emissions or resources used (e.g., metric tons of air pollutants emitted, or gallons of water used)
- Quantify the positive or negative impact (e.g., decreased biodiversity due to water scarcity or decreased life expectancy due to air pollution)
- Value in monetary terms the costs of the impact to society

Businesses typically use tools such as carbon or water footprints, life-cycle analysis, impact pathway analysis, or material flow analysis to quantify the amount of resources used and their impact on society or the environment. Each impact can have several consequences. For example, water depletion can affect society (loss of drinking water or decreased food supply), as well as the environment (not enough water to sustain fish and wildlife), and business itself (lack of clean fresh water so the business needs to install treatment facilities). A variety of standard economic techniques, such as abatement costs, contingent valuations, avoided costs, revealed preferences, are used to value natural capital costs depending on the data available and specific application or business decision.

Many environmental and social impacts are site specific because they depend on local conditions. The ideal situation is to use site-specific primary data, however, many times this is not possible given budgets, schedules, or technical constraints. As an alternative, scientists may use the **value transfer method**. In this method, the economic value is estimated by transferring available information from completed studies to another location or context by adjusting for known variables (such as population density, income, or ecosystem type and size).

CASE STUDY EXAMPLES

Over the years, the number of organizations publicly disclosing their involvement in natural capital valuation initiatives continues to grow and expand. In 2022, more than 330 companies encompassing revenues of over $1.5 trillion called for mandatory nature assessment and disclosure for all businesses and financial institutions at the **UN Biodiversity Conference**. Multinational companies like BNP Paribas, Danone, H&M Group, and IKEA, among others, recognize the need to be accountable and responsible in their part of nature recovery (Capitals Coalition 2022). The following case study examples highlight companies measuring and managing how they use and depend on natural capital, which in turn informs how they make and market greener brands (Capitals Coalition 2022).

NATURA

Driven by a desire to understand the full value it delivers to society, the Brazilian multinational cosmetics company, **Natura & Co, implemented an integrated profit and loss report** quantifying its positive and negative natural, social, human, and financial impacts in monetary terms (see Figure 6.5). Its approach considered

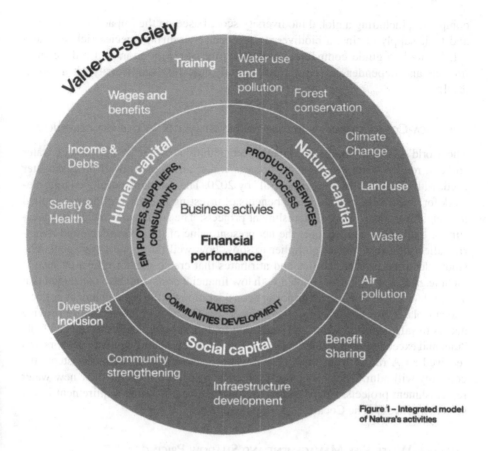

Figure 1 – Integrated model of Natura's activities

FIGURE 6.5   Natura's Profit and Loss report.

the impacts across all categories of stakeholders—including employees, customers, and society at large. Their approach assumes value is degraded when pollutants are discharged, waste is sent to a landfill, or natural land is converted. Value is created when society receives free benefits, such as ensuring employees are all paid a living wage. In 2021, **Natura's integrated profit and loss report disclosed a net positive societal value of about $18 billion**, composed of mainly social and human capital. "For each $1 in revenue, Natura generated $1.5 in positive social and natural impact." Yet, the company still faces a cost of $0.5 billion in natural capital, attributable to end consumer use (Segatelli 2022).

ABN AMRO

ABN AMRO, the large Dutch bank, along with Bank Paribas Asset Management, AXA Investment Management, Sycomore Asset Management, and Mirova, launched a joint initiative to develop a pioneering tool for measuring investment impact on biodiversity. The approach includes three tools applied to equity portfolio companies, including a global biodiversity score based on the impact of companies and their supply chains, a biodiversity footprint measuring species richness, and a LIFE tool to guide companies to identify, measure, monitor, and reduce their impacts and dependencies on natural capital, including biodiversity (at a global level).

THE COCA-COLA COMPANY—RETURN ON INVESTMENT IN WATER CONSERVATION

The world's largest beverage company set ambitious water goals: protect water sources, reduce water use, treat all process water, and replenish all process water used, with the goal to be "water neutral" by 2020. The company developed a framework for objectively evaluating the benefits of investing in conservation and applied it to eight European water replenishment projects. The framework included a measure of total ecosystem change, the net present value of the investment, and a metric called total investment multiplier to show how well the money was spent. The study identified project features and attributes that create greater return on investment (e.g., a large footprint coupled with low financial costs) and also concluded that stakeholder values are also an important consideration for the replenishment projects (Coca Cola 2016). Since setting their objectives, Coca-Cola has helped to provide access to safe drinking water and sanitation to 248,819 people. In 2021, Coca-Cola "met and exceeded" their 2020 replenish goal for the past consecutive six years and received an A rating from the CDP on water disclosure. For their next steps, the company will administer benchmarks on carbon sequestration for their new water replenishment projects, where carbon credits may be a possible requirement in the selection process (The Coca-Cola Company 2022).

ECOLAB—WATER RISK MANAGEMENT AND SHADOW PRICING

Ecolab, the $14 billion global leader in water, hygiene, and energy technologies and services, wanted to help companies understand the financial risks they are facing from physical water scarcity and quality risks resulting from pollution. The Water Risk Monetizer was launched in 2014 to help companies assess the financial implications of local market failures to price water according to its availability. In 2020, Ecolab updated the tool to include the World Resources Institute's Aqueduct Water Risk Atlas newly released water stress dataset, adjust global water pricing data to meet present demands, and updated global GDP population model (Ecolab 2020).

A unique aspect of the tool is the way it incorporates a management framework for acting on both water quality and quantity risks, based on the potential for increased operating costs or revenue at risk. The tool also applies a context-based water reduction target, by estimating the amount of water available to a business—its "share" of total water available to all businesses in a water basin—based on the facility's contribution to the local economy (Figure 6.6). Because water is a shared resource among many users, it is essential for a business to understand how its allocation may change, and with it any potential for constraints to continued revenue growth. Companies use the tool to stress test business models, identify high-risk assets or regions, and build the business case for water efficiency where it most matters.

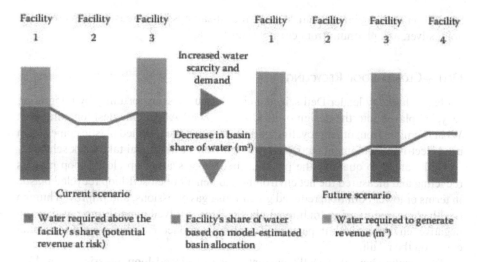

FIGURE 6.6 Ecolab business value with shadow water prices.

Green Electronics Council—Circular Economy and Greener Product Development

The Green Electronics Council used natural capital valuation tools to assess the business case for circular economy practices throughout the electronics sector. The results (see Figure 6.7) demonstrated that the industry could achieve a further

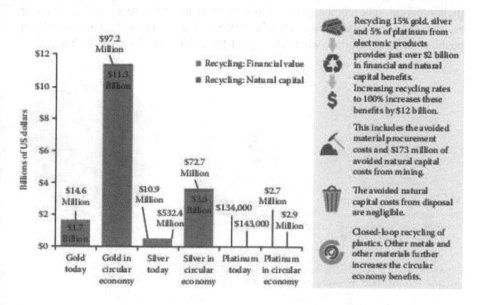

FIGURE 6.7 The benefit of greener electronics and a circular economy.

$10 billion in financial and natural capital cost-savings by increasing its recovery of gold, silver, and platinum from current rates to 100%.

Dell—Closed-Loop Recycling

Global technology leader Dell seized a circular business opportunity by integrating recycled plastic into the design of its OptiPlex 3030 All-in-One desktop computer. What is unique from other recycling initiatives is that the recycled plastic comes from used electronic equipment recovered through Dell's own global take-back scheme.

Dell wanted to quantify the potential benefit of scaling up closed-loop plastics recycling and measured the net environmental benefit of closed-loop recycled plastic in terms of lower pollution, reduced greenhouse gas emissions, and improved human health compared to using traditional plastic. This involved quantifying positive and negative environmental impacts and putting a monetary value—natural capital cost—on the result.

The results show that Dell's current usage of **closed-loop plastic has a 44% greater environmental benefit compared to virgin plastic**, equivalent to an annual saving of $1.3 million in avoided environmental costs. Of critical importance are the reduced human health and eco-toxicity impacts achieved by closed-loop recycling of plastic instead of disposal.

If all of Dell's plastic was supplied by closed-loop recycling, the environmental benefit to society would increase to $50 million per year (see Figure 6.8). If the entire computer manufacturing industry switched to using closed-loop recycled plastic, the environmental benefit would increase to $700 million per year (Dell 2015).

Dell has collected more than 100 million pounds of recycled plastic parts through their program. The company has also expanded its initiative to include e-waste, recovering over 2.1 billion pounds of e-waste (Ward 2021). Dell continues to keep pushing for sustainability, where their goal for 2030 is to "have more than half of

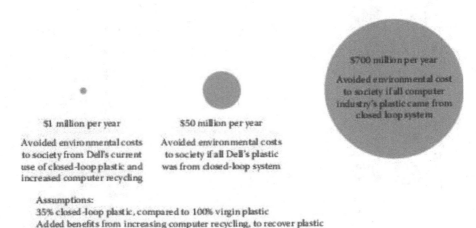

$700 million per year

Avoided environmental cost to society if all computer industry's plastic came from closed loop system

$1 million per year

Avoided environmental costs to society from Dell's current use of closed-loop plastic and increased computer recycling

$50 million per year

Avoided environmental costs to society if all Dell's plastic was from closed-loop system

Assumptions:
35% closed-loop plastic, compared to 100% virgin plastic
Added benefits from increasing computer recycling, to recover plastic

FIGURE 6.8   Net benefit of closed-loop recycling. (Courtesy of Trucost, London, UK.)

their product content made from recycled, renewable or reduced carbon emissions material" (Dell 2023).

Sveaskog—Integrated P&L Statement

Sveaskog, the owner of 14% of Sweden's forests, commissioned GIST Impact to produce an Integrated Profit and Loss statement. The analysis **estimated that its net positive natural capital impacts to be around seven times the firm's annual profits**, largely because of the significant positive impact of carbon sequestration and water conservation services generated by the firm's forest assets.

Kering—Greener Product Design Tools

Kering, the French luxury goods holding company, which owns Alexander McQueen, Balenciaga, Gucci, Puma, Volcom, and other sport and lifestyle brands, developed the revolutionary concept for an **Environmental Profit and Loss framework** in 2011 and has since applied it across the operations and supply chain of its enterprise. In 2016, the company in collaboration with the Parsons School of Design created the My EP&L app that incorporates the EP&L framework into a simple, easy-to-use tool to measure and better understand the environmental impacts of students' creations. For example, My EP&L shows that by choosing a bag made from French leather with the inner lining in Chinese silk and hardware in brass from Chile instead of a bag made out of US leather with the inner lining in Chinese linen and hardware in Chinese bamboo, there is €4.40 less EP&L impact or a 26% environmental saving (Kering 2016).

Kering set its target to reduce their EP&L footprint by 40% across the supply chain 2025 in comparison to 2015. In 2021, the company achieved its goal of decreasing the EP&L footprint by 41%, four years ahead target. Over the next five years, Kering's objective focuses on "transforming 1,000,000 hectares of crop and rangelands into regenerative agricultural spaces" in collaboration with Conservation International to bring in positive value (Kering 2021).

CONCLUSION

Intensifying climate change and biodiversity loss are causing companies and investors to question—and understand—how their future viability depends on nature. Many efforts have been voluntary. But policymaking—whether in the form of regulation or carbon taxation—is increasing the speed and breadth of companies' and investors' actions.

The TNFD and Science Based Target Initiative (SBTi) are emerging frameworks for understanding, measuring, and reporting on biodiversity loss. In addition, natural capital valuation tools are being used to quantify environmental risks in business terms. **Quantifying the unpaid cost of environmental resource impacts and dependencies from pollution to natural resources and integrating location-specific shadow prices in decision-making provide vital insights that can help companies manage natural capital costs that are not yet fully priced by the market.**

A growing number of companies are adopting this approach and putting shadow prices on carbon and other pollution emissions, water use, land conversion, and other natural resources to get ahead of the trend toward regulators, society, and nature correcting market failures to price-constrained environmental resources.

For many, it is a risk-management strategy—a way to identify high-impact environmental issues, assets, regions, and value-chain segments, and talk about risk in the language of business: dollars and cents. For others, it is a way to build the business case for energy efficiency, water conservation, and sustainable resource management where it most matters.

As a brand owner, **what are tangible next steps you can take to begin to understand how your brand impacts and depends on nature and biodiversity**?

1. Understand how your operations, suppliers, products, and raw materials rely on natural capital and biodiversity. A simple "heat map" visualizing your company's value chain and where there are impacts and dependencies on nature can be a powerful tool for educating your colleagues who are not sustainability experts. For example, a personal care products company might highlight naturally sourced raw materials from regions that are at risk from biodiversity loss as well the carbon pollution caused by heating water when the products are used.

2. Understand which of those impacts and dependencies are most likely to create financial risks (or opportunities) for your brand, either because of regulatory risks, operational or supply-chain production risks, or loss of a license to operate in communities where natural resources are constrained.

REFERENCES

American Chemistry Council. 2016. Plastics and Sustainability: A Valuation of Environmental Benefits, Costs, and Opportunities for Continuous Improvement. https://plastics. americanchemistry.com/Study-from-Trucost-Finds-Plastics-Reduce-Environmental-Costs/ (Accessed December 29, 2016).

American Forest and Paper Association. 2016. http://www.afandpa.org/our-industry/ economic-impact (Accessed December 29, 2016).

Coca Cola. 2016. Natural Capital Accounting, the Coca Cola Water Replenishment Project. http://naturalcapitalcoalition.org/wp-content/uploads/2016/07/Denkstatt_Natural_Capital_Accounting.pdf (Accessed December 29, 2016).

Dell. 2015. Valuing the Net Benefit of Dell's More Sustainable Plastic Use. http://i.dell.com/ sites/content/corporate/corp-comm/en/Documents/circular-economy-net-benefits.pdf (Accessed December 29, 2016).

Ecolab's Water Risk Monetizer Updates Global Water Data to Reflect Current Trends. 2020. Ecolab. https://www.ecolab.com/news/2020/01/ecolab-s-water-risk-monetizer-updates-global-water-data-to-reflect-current-trends (Accessed August 21, 2023).

Environmental Profit & Loss 2021 Group Results. 2021. Kering. https://keringcorporate.dam. kering.com/m/5edba9133d460b06/original/Kering-Environmental-Profit-and-Loss-Report-2021-EN-Only.pdf (Accessed August 21, 2023).

How to create Natural Capital through Nature-Based Solutions Progress Report. 2022. The Coca-Cola Company. https://www.coca-cola.eu/news/supporting-environment/creating-natural-capital-through-nature-based-solutions-2022 (Accessed August 21, 2023).

Kering Group EP&L. 2015. https://www.coca-cola.com/eu/en/media-center/creating-natural-capital-through-nature-based-solutions-2022 (Accessed December 29, 2016).

Kering, MyEP&L App. 2016 https://www.kering.com/en/sustainability/measuring-our-impact/our-ep-l/ (Accessed April 24, 2024).

More Than 330 Businesses Call on Heads of State to Make Nature Assessment and Disclosure Mandatory at COP15. 2022. Capitals Coalition. https://capitalscoalition.org/make-it-mandatory-cop15/ (Accessed August 21, 2023)

Natural Capital at Risk—The Top 100 Externalities of Business. 2013. TEEB. http://naturalcapitalcoalition.org/wp-content/uploads/2016/08/Trucost-Nat-Cap-at-Risk-Final-Report-web.pdf (Accessed December 29, 2016).

Nature Risk Rising: Why the Crisis Engulfing Nature Matters for Business and the Economy. 2020. The World Economic Forum. https://www3.weforum.org/docs/WEF_New_Nature_Economy_Report_2020.pdf (Accessed August 17, 2023).

Our ESG Goals. 2023. Dell Technologies. https://www.dell.com/en-us/dt/corporate/social-impact/reporting/goals.htm (Accessed August 21, 2023).

Puma Environment. 2020. Puma. https://annual-report.puma.com/2020/en/sustainability/environment.html (Accessed August 20, 2023).

Puma EP&L. 2024 https://www.spglobal.com/marketintelligence/en/news-insights/research/puma-environmental-profit-loss-account (Accessed April 24, 2024).

Segatelli, A. 2022. Case Study: Integrated Profit and Loss Accounting at Natura &Co. International Federation of Accountants (IFAC). https://www.ifac.org/knowledge-gateway/supporting-international-standards/discussion/case-study-integrated-profit-and-loss-accounting-natura-co (Accessed August 20, 2023)

S&P Global. 2023. Solving for interconnected sustainability challenges. https://www.spglobal.com/esg/insights/featured/special-editorial/how-the-world-s-largest-companies-depend-on-nature-and-biodiversity

Stewart, L. 2022. Asset Managers Start Adopting Policies Around Biodiversity. Morningstar. https://www.morningstar.com/sustainable-investing/asset-managers-start-adopting-policies-around-biodiversity. (Accessed April 23rd, 2024).

TD Bank. 2015. Natural Capital Valuations. https://www.td.com/document/PDF/corporateresponsibility/2015-Natural-Capital-Valuations.pdf (Accessed December 29, 2016).

The Economics of Biodiversity: The Dasgupta Review. 2021. https://www.gov.uk/government/publications/final-report-the-economics-of-biodiversity-the-dasgupta-review. Accessed April 24, 2024.

The Future of Nature and Business. 2020. The World Economic Forum. https://www3.weforum.org/docs/WEF_The_Future_Of_Nature_And_Business_2020.pdf (Accessed August 18, 2023).

Ward, A. 2021. What Goes Around, Comes Full Circle. Dell Technologies. https://www.dell.com/en-us/blog/what-goes-around-comes-full-circle/ (Accessed August 21, 2023).

Wealth Accounting Valuation System (WAVES). 2016. https://www.wavespartnership.org/ (Accessed December 29, 2016).

Section III

Marketing Greener Products

7 Green Marketing

Al Iannuzzi

Green marketing is now a necessity rather than an initiative to be taken by proactive companies or those with a niche green product line. We are seeing green marketing occurring in all types of products and industries. Global consumers are demanding products that have greener attributes. A recent Deloitte study indicated that "67% of participants were willing to pay up to 41% more" for sustainable products and that percentage increases among younger age groups.

Companies are profiting from marketing more sustainable products. Green Giants, or large-scale firms that prioritize sustainability, have over $1 billion in sales of more sustainable products. Companies have asked their suppliers to provide them with greener products; Kaiser Permanente and Procter & Gamble (P&G) have developed scorecards to judge the sustainability of supplier's products to facilitate decisions for more sustainable products. Various examples of green marketing are discussed via case study analysis, including companies such as Honest Tea, Timberland, GE Ecomagination, and BASF.

THE CASE FOR GREEN MARKETING

Green marketing is now a mainstream necessity rather than an initiative to be taken by proactive companies or those with a niche green product line. We are seeing green marketing occurring in all types of products and industries. It's not just the firms that sell direct to consumers must be concerned with this new marketing; suppliers of all types of products must start heeding this expectation. Companies are seeing a need to develop greener products in all parts of business; automotive, consumer products, chemicals, electrical products, medical products, and even pharmaceuticals.

Working for a company that sells products directly to consumers and previously working for one that did a good deal of business-to-business (B2B) sales, my experience tells me that green marketing has become a business imperative. I have seen green marketing take off in the consumer area, but I am now seeing an uptick in the B2B space too. Customers in all markets are demanding greener products. The signals in the field are building, and now is the time to start positioning your brand to get on board with this demand.

CONSUMER DEMAND

Global consumers are demanding products that have greener attributes. A study from the consulting firm Deloitte indicates that "67% of participants were willing to pay up to 41% more" for sustainable products and that percentage increases among younger age groups. Moreover, a joint study between McKinsey and NielsenIQ

DOI: 10.1201/9781003441939-10

found that "**products making ESG-related claims accounted for 56% of all retail sales growth**" between 2018 and 2022 (McKinsey 2023).

"67% of participants were willing to pay up to 41% more" for sustainable products, and that percentage increases among younger age groups

Grundmann et al. (2022)

We have seen greener products entering the mainstream in a major way. In the United States, local supermarkets now sell products that were once exclusive to health food stores. Products like Seventh Generation detergents, Method facial cleansers, Ms. Meyers had soap, and Stonyfield Farm organic yogurt, to name a few, are commonplace. It is typical for grocery stores to have an "organic" section offering cereals, eggs, fruits, vegetables, milk, and baby products. On top of this, we see traditional companies coming out with greener products that have done extremely well. Clorox's Green Works line is perhaps the best example of this. According to Clorox's own press release, one year after introducing Green Works, the natural cleaning category grew more than 100%, delivering on the company's goal to make natural cleaning more accessible and affordable to everyday consumers. "Green Works® was the #1 natural cleaning brand in the United States upon release, capturing a 42 percent market share" (The Clorox Company 2010).

SHOULD YOU MARKET GREEN?

What good is a greener product if no one knows about it?

The growth of the organic industry is another proof point for consumers' growing demand for healthier options. Organic sales have doubled in the last decade; the organic industry grew 4% in 2022 totaling a record $67.6 billion (Organic Trade Association 2023). It is estimated that organics now make up almost 6% of total food sales in the United States. This growth is not just limited to food. Non-food items, such as organic clothing, supplements, and personal care products, totaled over $5.9 billion in sales. Many large companies are jumping on this opportunity and acting in response to these global trends.

10 BIG BUSINESS RESPONSES TO MEET CONSUMER DEMAND FOR GREENER PRODUCTS

1. *Panera Bread* "No No List": a list of more than 150 artificial preservatives, sweeteners, colors, and flavors that it plans to remove from its menu.
2. *Chipotle* becomes the first fast food chain to vow to remove genetically modified food (GMOs) from its menu.

3. *McDonald's* announced it will only buy and sell "chicken raised without antibiotics that are important to human medicine." It also announced it will offer milk from cows that are not treated with the artificial growth hormone recombinant bovine growth hormone in their US restaurants later this year.

4. *Tyson Foods*, the largest US poultry producer, announced plans to achieve carbon neutrality by 2035.

5. Ice cream giant *Breyers* pledges to stop using milk from cows treated with recombinant bovine growth hormone, which has been purported to be linked to several health problems in humans.

6. *Shake Shack*, the owner of Panera, Five Guys, and Chipotle, announces the value of its initial public offering had increased to $675 million. They have a focus on healthier, ethical, and more sustainable food experiences.

7. *Home Depot and Lowe's*, the largest home improvement retailers, will phase out phthalates from vinyl flooring.

8. *Lowe's* commits to stop selling pesticides alleged to be killing bees (neonicotinoid) to protect pollinators.

9. *Adidas* announced plans to turn ocean plastic into sportswear materials made from ocean plastic waste.

10. *Levi's* and other companies commit to water conservation in drought-stricken California (EcoWatch 2015).

SUSTAINABLE BRANDS ARE MORE PROFITABLE

In her book Green Giants, Freya Williams, co-founder of sustainability communication firm OgilvyEarth, cites data that leading companies in sustainable, social, and governance have **25% higher stock value** than their competitors and **grow at double the rate than other brands.** She goes on to make the case that greener products are profitable and highlights nine companies with **over $1 billion in sales** that are attributed to more sustainable products. As we can see from the list of these companies, greener products cut across various sectors and are not only better for people and the planet but also are yielding significant profit. This is another reason why every company should be focusing on making and marketing greener products.

The list below indicates nine companies that have businesses with over a billion US dollars in annual revenue connected to a product or service with "sustainability at its core." It should be noted that this is not a comprehensive list because there are other companies that have over a billion in revenue from greener products that are not included here. For one, the company I once worked for, Johnson & Johnson, had over $8 billion in Earthwards® recognized greener products on the market.

Companies with over $1 Billion in Sales of Greener Products

Brand	Sector	2014 Revenue in Billions ($)
Nike Flyknit	Athletic Apparel	1
IKEA (products for a more sustainable life)	Furniture	1.13
Natura	Consumer products	2.65

Brand	Sector	2014 Revenue in Billions ($)
Tesla	Automotive	3.2
Chipotle	Food	4.11
Whole Foods Market	Food	14.19
Toyota Prius	Automotive	15.44
GE Ecomagination	Diverse products	28
Unilever	Consumer products	52.37

Source: Williams. Green Giants (2015).

GREENER PRODUCTS SHOULD NOT COMMAND A HIGHER PRICE

This idea of **greener being an** *"and"* is a concept I believe is proven out by numerous studies on consumers purchasing behavior. In other words, most consumers want a greener product, but don't necessarily want to pay more for it, so it is an *and*. Only the deep green purchasers, or the ones the Shelton Group would call "Actives," may be willing to pay a premium for a greener product (34% of shoppers in the United States) (Shelton 2022). **I have seen numerous studies that indicate the desire of consumers to pay more for greener products but when it comes to the point of sale, other aspects win out like quality and price.** Therefore, to reach most purchasers, product attributes like efficacy and appeal must be present, and then its sustainability qualities can push it over the edge. I like to say that all things being equal with your competitor (price, quality etc.) and **if you can demonstrate that your product is more sustainable, you should win in the marketplace**. Consumer product buyers are not the only ones looking for greener products; we are seeing companies having to market their products' greener aspects to other businesses to get market advantage too. Therefore, the concept of the greener characteristics being an *"and"* also applies to the business customer as well.

> This idea of **greener being an** *"and"* is a concept I believe is proven out by numerous studies on consumers purchasing behavior. In other words, most consumers want a greener product, but don't necessarily want to pay more for it, so it is an *"and."*

BUSINESS-TO-BUSINESS DEMAND

Like consumers, many companies are actively seeking to purchase greener products. Firms have been pressured by market forces to take responsibility for their supply chain. Perhaps the most notable example of this is when Nike and other footwear and apparel companies had received significant pressure to take responsibility for the poor working and environmental conditions at the companies that manufactured shoes for them in the 1990s. Not long after this, several apparel firms had received significant pressure to take responsibility for suppliers that made their products in sweatshops. Companies started to get the point and then began to develop very comprehensive policies and auditing programs for their suppliers to ensure that the environment is protected and employees within their supply chain have safe working conditions. Taking this initiative even further, we see manufacturing firms now asking their supply chain to achieve certain goals like reducing water, energy, waste, and other issues. These initiatives have been coined as "**greening the supply chain.**"

With the advent of greening the supply-chain initiatives, suppliers will gain market advantage when they meet the demands of their business customers. I know that in my company, we give preference to like-minded firms that have adopted more sustainable practices, especially when it comes to Green House Gas GHG reductions and green power adoption. Meeting your customer's needs is the inspiration behind B2B marketing.

As an illustration, look at one of Unilever's sustainability goals: to increase the amount of sustainably sourced raw materials such as palm oil or tea (Unilever 2023). A supplier that can reliably source sustainable tea can assist them in achieving this goal while creating a sales growth opportunity.

Similarly, suppliers to Tesco, the largest retailer in the United Kingdom, can see that they are very serious about getting sustainable innovation from their suppliers. They have a comprehensive Environmental Guide for Suppliers and established targets to drive improved performance. Being a major global retailer, Tesco recognizes that they have an opportunity to make a big impact in their supply chain by conducting their operations in a responsible way. Strides have been made to reduce their carbon footprint, and they want to encourage their supply base to do the same. A goal was established to become a zero-carbon business by 2050 and to reduce CO_2 emissions 30% from the products in their supply chain against a 2008 baseline by 2020. If you sell to Tesco, you will have to put plans in place to address CO_2 since this goal affects all suppliers.

Addressing carbon emissions isn't the only concern for suppliers; Tesco identified climate change, water use, and biodiversity as key areas for their suppliers to focus on. They believe that in making strides to reduce their footprint, targets need to be established for their biggest agricultural products by 2025, specifically:

- Thirty-percent reduction in greenhouse gas emissions.
- Reduction in water use, including local reduction targets for water-stressed areas.
- Improvement in farmland biodiversity (soil health, pollinators, and off-field biodiversity).

Tesco goes even further for certain suppliers of agricultural commodities because poor practices from farming have been associated with deforestation. A goal of Net Zero Carbon was set for 2050 (Tesco 2021). The key commodities being focused on are palm oil, cattle products, soy, and timber. In addition, mandatory policies for commodities of high concern have been established for suppliers to follow, for example, palm oil sourcing and sustainable sea food. Any company that can excel in these areas will at a minimum get the right to sell in a Tesco store and, if they have superior performance, could get the most preferred shelf space or other marketing benefits.

It's not just Tesco that has green purchasing policies. You would be hard-pressed to find a large company that doesn't have a procurement strategy that emphasizes purchasing greener products. Focus has been brought on suppliers in various product categories like energy savings, and electronics, to paper that contain post-consumer recycled (PCR) content or greening the supply-chain initiatives that seek suppliers to produce sustainable innovations. These kinds of initiatives will only become broader and demand more eco-efficiency from suppliers in the future. **The companies that see this trend and can meet customers' demands will be able to win in the greener product market.**

SUPPLIER SCORECARDS

B2B marketing also has a big impact and one of the reasons is the advent of supplier scorecards. Walmart's Supplier Sustainability Assessment and Packaging Score Card is well known, but there are other very influential companies with sustainability scorecards of their own. Large companies like Kaiser Permanente and P&G have also issued mandatory supplier scorecards. At Estée Lauder Companies I have seen a growth in retailers like Selfridges, Sephora, Ulta, and Macy's adopting sustainability information on topics such as ingredients, packaging, and the use of third-party certifications. Suppliers take notice when they hear the Vice President and Chief Procurement Officer of a major health-care provider, Kaiser Permanente, say "green up your act today, lest you lose a huge client tomorrow." Also, consider that Kaiser's purchasing comes to about $14 billion a year and P&G alone has 75,000 suppliers throughout the world (Guevarra 2010).

These scorecards will no doubt drive companies to highlight the greener benefits of their products. Clear communication of eco-friendly features in your product portfolio is crucial when customers are interested in whether or not the products, they are buying contain specific toxic chemicals, or if they are manufactured using renewable power.

EXAMPLES OF QUESTIONS FROM P&G AND KP SCORECARDS

- What % of energy consumed is generated from renewable resources?
- Does the company have a climate action plan with baseline and targets?
- How many metric tons of hazardous and nonhazardous waste are produced?
- Free of intentionally added bisphenol A or bisphenol A derived chemicals (including thermal paper)?
- Free of polyvinyl chloride?
- Primary Packaging—Contains more than 10% PCR content? (Fast Company 2010)

Other drivers of B2B marketing are the public relations benefits for being perceived as a company that cares, and helps customers save money. Home-improvement giant Lowe's was recognized as the WaterSense® retail partner of the year by the US Environmental Protection Agency. They also received kudos from the Department of Energy's ENERGY STAR® Sustained Excellence Award. These accolades were earned by providing cost-saving products that also minimize environmental impact. Lowe's explains the benefits of water conservation in their stores and helps families reduce utility bills. Their ENERGY STAR products saved customers $23.5 billion in energy costs between 2018 and 2021 (Lowe's 2021).

With this kind of focus on greener products it makes sense for suppliers to get the ENERGY STAR or WaterSense labels on their products if they want to sell to Lowe's. With Lowe's getting such positive recognition as a sustainable company while generating savings for their customers, it would be prudent for suppliers to provide products that aligns with Lowe's goals.

This scenario is common across all industrial sectors. Companies that provide building supply products can help customers seeking green building certifications like Leadership in Energy and Environmental Design (LEED) with products such as low volatile organic compound (VOC) paint, more energy-efficient windows, sustainable wood products, more efficient air-handling equipment, more sustainable carpet and cleaning compounds, to name a few. With suppliers being held to higher standards, like sustainable innovation, providing greener products to business customers helps them achieve sustainability goals while increasing sales. It's a win-win.

EXAMPLES OF GREEN MARKETING

Earlier, we discussed the practices leading companies use to make greener products. **Step one in green marketing is to have a credible greener product** to bring to customers. The next step, which is equally important, is effectively communicating how the qualities of your product or service meet consumer demands. Telling your story is super important, I like to say what good is a greener product if no one knows about it? **Just as important as making a product more sustainable is the communication of its benefits.**

I like to say what good is a greener product if no one knows about it?

What are the key elements of green marketing? **First, your product must have a "greener" story to tell, based on scientific facts and data**. You must also understand the market segment you are selling into and seek to meet your customer's greener product demands. Finally, the greener benefits must be appropriately communicated, without overstating or misleading.

The Boston Center for Corporate Citizenship prescribes **five guidelines for green marketing**.

1. *Be precise*: Make specific claims that provide quantitative impacts.
2. *Be relevant*: Demonstrate a clear connection between the product or service and the environment.
3. *Be a resource*: Provide additional information for consumers in a place where they want it.
4. *Be consistent*: Don't let marketing images send a signal that contradicts the carefully chosen words and facts you use.

5. *Be realistic*: Continuous environmental improvements are a crucial aspect of any product or service's broader environmental journey (Hollender et al. 2010). The way I would say this is, **communicate your products as green*er*, not green.**

Based on my experience, I would simplify effective green marketing into three key elements:

(1) Having a truly greener product to market (see Chapter 4).
(2) Understanding customer requirements.
(3) Appropriately communicating greener characteristics.

KEYS TO GREEN MARKETING

- Have a credible greener product story.
- Meet your customers' greener product demands.
- Appropriately communicate the product's greener attributes.

BUSINESS-TO-CONSUMER (B2C) EXAMPLES

Green marketing is all about communication. **Making the greenest product is useless if no one knows about it**. Let's examine some successful green marketing campaigns through the key elements we just discussed.

CLOROX GREEN WORKS

Clorox Green Works changed the game and enabled greener products to be widely available. Before Green Works, green products were often stigmatized as being a niche product sold primarily in health food stores. Consumers perceived them as ineffective, expensive, and untrustworthy. Green Works is based on a natural ingredients platform (at least 99% natural ingredients), proven to perform just as well as and sometimes better than existing cleaning products. Consumer research conducted by NYU Stern revealed that **consumers care most about protecting themselves and their families** (NYU Stern 2023).

To establish authenticity, Clorox pursed the US EPA's Design for the Environment (DfE) certification (now called Safer Choice). Clorox further bolstered their green claims by receiving endorsement from The Sierra Club, displaying their logo on Green Works product bottles. Clorox also placed the ingredients of the compounds used on the product labels, even though this was not required by law and is not a common practice for household cleaners (Werbach 2009).

Green Works clearly has done the work to credibly develop a greener product. They understand consumers' needs for personal protection, and they have clearly messaged the products greener attributes to the customer. Receiving third-party endorsement with the DfE certification and the Sierra Club's backing helped cement the product as truly green in the mind of prospective purchasers.

SEVENTH GENERATION

Seventh Generation, a deep green company from the beginning, sells a diverse range of products: cleaning products, baby, laundry, dish detergents, and more. Their mission to bring greener products into the mainstream market is inspired by an Iroquois Native American law on considering the impact of their decisions on the next seven generations.

Seventh Generation effectively communicates the greener nature of their products. Their website contains a wealth of sustainability information, showcasing products based on natural ingredients and featuring eco-labels such as USDA Certified Biobased, cruelty-free (Leaping Bunny), How to Recycle logo, and even gluten-free options. They share sourcing policies, indicating supplier requirements, and even the location of manufacturing sites—important issues for conscious consumers. Moreover, they have an ingredients' "glossary" that explains the purpose and environmental impact of every material used in their products. Customers can use this to look up what an ingredient is used for and if it is safe for the environment. As an example:

Ingredient: *Aloe barbadensis* leaf juice.
Use: Skin onditioner.
Environmental impact: Plant derived, biodegradable.

Let's look at the messaging for the Hand Wash: "Hand soap made with plant-based ingredients, formulated with you and your family in mind. Made for sensitive skin and made without fragrances, dyes, or triclosan. Our hand wash bottle is made with 100% recyclable plastic." Eco-labels used are USDA Certified Biobased, Leaping Bunny certified, hypoallergenic, EPA Safer Choice certified, hypoallergenic, fragrance free, recyclable bottle. A full disclosure of all ingredients is available on the website.

COMMUNICATING GREEN ATTRIBUTES IS IMPORTANT TO THE CUSTOMER

Seventh Generation Hand Wash—Free & Clean	
Messaging	
Formulated with you and your family in mind	Made without triclosan, dyes, or phthalates
Made for sensitive skin	Fragrance free, hypoallergenic Free & Clean formula
Plant-based	Made with aloe leaf juice + Other plant-derived ingredients, Leaping Bunny certified

Source: Seventh Generation, 2023, Hand Wash—Free & Clean, https://www.seventhgeneration.com/hand-wash-free-clean-unscented.

What's Inside Our Safe and Effective Formula

Water, sodium coco-sulfate (plant-derived cleaning agent), decyl glucoside (plant-derived surfactant), aloe barbadensis leaf juice (plant-derived feel enhancer), sodium chloride (mineral-based viscosity modifier), magnesium chloride (mineral-based viscosity modifier), glycerin (plant-derived foam enhancer), glyceryl caprylate/caprate (plant-derived skin conditioner), citric acid (plant-derived pH adjuster), tetrasodium iminodisuccinate (synthetic preservative enhancer), coco-glucoside (plant-derived skin conditioner), glyceryl oleate (plant-derived skin conditioner), and sodium benzoate and potassium sorbate (synthetic preservatives).

Product Manufactured in: South Carolina, USA.
Ingredient Origins: USA and Globally Sourced.

(Seventh Generation 2023)

Seventh Generation covers all the key areas for green marketing. They have a truly greener product, they know what customers are looking for (more natural ingredients), and they clearly communicate the greener benefits to meet customer demands.

Honest Tea

I discovered Honest Tea through my teenage children, who desired a low-calorie, tasty drink. The name reflects the mission: "Honest Tea creates and promotes delicious, truly healthy, organic beverages. We strive to grow with the same honesty we use to craft our products, with sustainability and great taste for all." Honest Tea is committed to social responsibility, striving for "authenticity, integrity and purity, in our products and in the way we do business." Their platform is a healthy beverage with a lot less sugar than most bottled drinks. In March 2011, The Coca-Cola Company purchased Honest Tea; however, it is run as an independent business unit.

There Are Five Pillars to Their Mission
- Promoting Health & Wellness.
- Reducing Our Environmental Footprint.
- Democratizing Organics.
- Creating Economic Opportunity.

(Honest Tea 2020)

Honest Tea uses independent laboratory analysis to prove that its drinks have antioxidant levels that are as high as or higher than brewed tea leaves. In 2003, Honest Tea became the first to make a Fair Trade Certified™ bottled tea. This certification strives to empower family farmers and workers around the world to get a fair price for their harvest, have safe working conditions, and earn a living wage (Honest Tea 2020).

Concerned with unhealthy super-sweet beverages loaded with sugar, Honest Tea offers beverages without any sugar, such as "Just" Green Tea (no calories, no sugar). They also produce products that are a "tad sweet"—drinks sweetened with organic cane sugar which contain 60 calories or fewer. For example, Heavenly Lemon Tulsi Herbal Tea is a brew made with lemon juice, tulsi, and rooibos leaves, and an aromatic blend of lemongrass, lemon peel, and lemon myrtle, and has only 60 calories.

Honest Tea beverages and tea bags are certified to the USDA's organic standards. The certification ensures customers that the products' raw materials were grown following organic farming techniques and do not contain antibiotics, pesticides, irradiation, or bioengineering. The farms that provide the raw materials are examined by third-party certification agencies.

HONEST TEA MESSAGING TO THE HEALTH-CONSCIOUS CONSUMER

All of our teas are certified organic according to standards set by the US Department of Agriculture (USDA) and enforced by accredited third-party certifying agents.

Typically, glass, PET plastic, boxes, and pouches for children's drink containers are used for packaging. While pouches are typically not recyclable, their aseptic Tetra Brik® package is recyclable in over 50% of US municipalities. Because most drink pouches also cannot be recycled in curb side programs, Honest Tea partnered with TerraCycle to upcycle these packages and uniquely converts them into useful items like fashion bags, tote bags, pencil cases, and other items. This type of innovative program gives customers confidence in Honest Tea because it conveys that they are doing their best to facilitate the reuse of their packaging.

Honest Tea clearly communicates the greener benefits of their products on bottle labels. As an example, the Fair Trade Certified and USDA Organic logos along with the calories are prominent on the label of their Peach OO-LA-Long "Just a Tad Sweet" tea. A visit to their website reveals further messaging that targets conscious consumer by using the following images: Fair Trade Certified, USDA Organic, Gluten Free, no GMOs, and 1% for the Planet. There is even a source map that depicts which parts of the world key ingredients such as tea, fruit, herbs, spices, and sweeteners are sourced from (Honest Tea 2020).

CERTIFICATIONS AND CLAIMS USED BY HONEST TEA

- Fair Trade Certified™
- USDA Organic
- Calories per bottle
- Gluten Free
- No GMOs
- 1% for the Planet

To further reinforce their commitment to sustainability, they have developed a "Mission Report." In the 2020 "Honest Mission Report," there are facts, figures, data, and details that build consumer confidence that they are keeping their eye on the sustainability ball. An example is the discussion of their donations to 1% for the Planet. They give 1% of sales from their 16 fl. oz. Glass-bottle tea line to organizations within the 1% for the Planet network. There is also an interesting discussion about their sourcing practices, where fair trade premiums are paid to suppliers for efforts to help make their practices more efficient so that they can garner more income.

Honest Tea has a good sustainability story to tell that is backed up with data and third-party certifications. Research indicates that the primary driver for organic food purchases is the desire for "better health." Consumers tend to favor endorsements or certifications for this category of products. Therefore, they are on target in addressing their customers' desires with the organic and fair-trade certifications they obtained and prominently display. Finally, they are credibly and very clearly communicating to their customers the products' greener traits.

TIMBERLAND

As pioneers of the Green Index®, Timberland has built sustainability into its core. Supplying footwear and apparel for the outdoors, they feel a direct connection to developing sustainable products—hence their environmental commitment is called "Earthkeeping." Timberland uses the Green Index label to depict their environmental improvements. Resembling a nutrition label, this label indicates the climate impact, chemicals used, and resource consumption. The label conveys criteria that provide consumers with a relative measure of a product's environmental impact to spur more sustainable purchasing.

An example of product improvements is the Earthkeepers® GT Scramble Lace Chukka.

This boot scored 4.5 out of 10 on the Green Index (10 being the worst score). Their three categories scored as follows:

- 3.0 for Greenhouse gases which are produced in making the raw materials during footwear production.
- 2.0 for Chemicals used in materials and footwear production.
- 7.5 for Resource Consumption for making the product. The idea is to use resources that use less land, water, and chemicals, and more recycled materials.

The product improvements are communicated through the Green Index label (Timberland 2015). Further demonstration that Timberland gets how to do green marketing right is the messaging for their Earthkeepers® original leather boot. Consumers looking for a boot that has been made with lower environmental impact can quickly see that thought was put into making this product greener (Timberland 2015).

MARKETING EARTHKEEPERS® ORIGINAL LEATHER BOOTS

- We haven't sacrificed quality or rugged good looks to make them eco-conscious.
- 100% recycled PET lining is made from recycled plastic bottles.
- 100% organic cotton laces.
- One or more major components use at least 50% leather tanned in a facility rated Silver or higher by a third-party environmental audit.

Timberland Earthkeepers® Original Leather Boots

Timberland has developed products using their Green Index (Figure 7.1) that have better environmental performance. Customer demand for transparency about the materials used in their products is being met. Consumer research has been used to identify the most important customer desires to communicate about.

BUSINESS-TO-BUSINESS (B2B) GREEN MARKETING

When we think of green marketing, our natural tendency is to think about direct-to-consumer products. However, **there is a far greater opportunity in B2B marketing**. Consider the millions of suppliers that are necessary for all the products in the marketplace. As mentioned above, P&G has 75,000 suppliers alone! I believe that in the B2B space, green marketing will be an imperative in the future, and for now it is a key product differentiator. So, let's evaluate some successful B2B green marketing campaigns and see how they compare to the B2C approach.

ECOMAGINATION

Perhaps one of the most successful B2B marketing initiatives is General Electirc's (GE) Ecomagination program. It would be hard to discuss green marketing without mentioning the inroads made by GE. There were Ecomagination television commercials, print advertisements, and digital marketing as well as an annual report that details the programs' success. GE's CEO is seen at numerous events throughout the world effectively communicating about this program.

Ecomagination was launched in 2005 and steadily grew into one of the most successful green marketing programs ever, paying dividends to GE and their customers. In fact, in 2015, sales of greener products resulted in $36 billion! (GE 2016). The whole reason for the existence of Ecomagination is to *meet customers' requirements*. Ecomagination was launched as a "groundbreaking strategy the company used to build more efficient machines that produce cleaner energy, reduce greenhouse gas emissions, clean water and cut its use, and make money while doing it" (GE 2015).

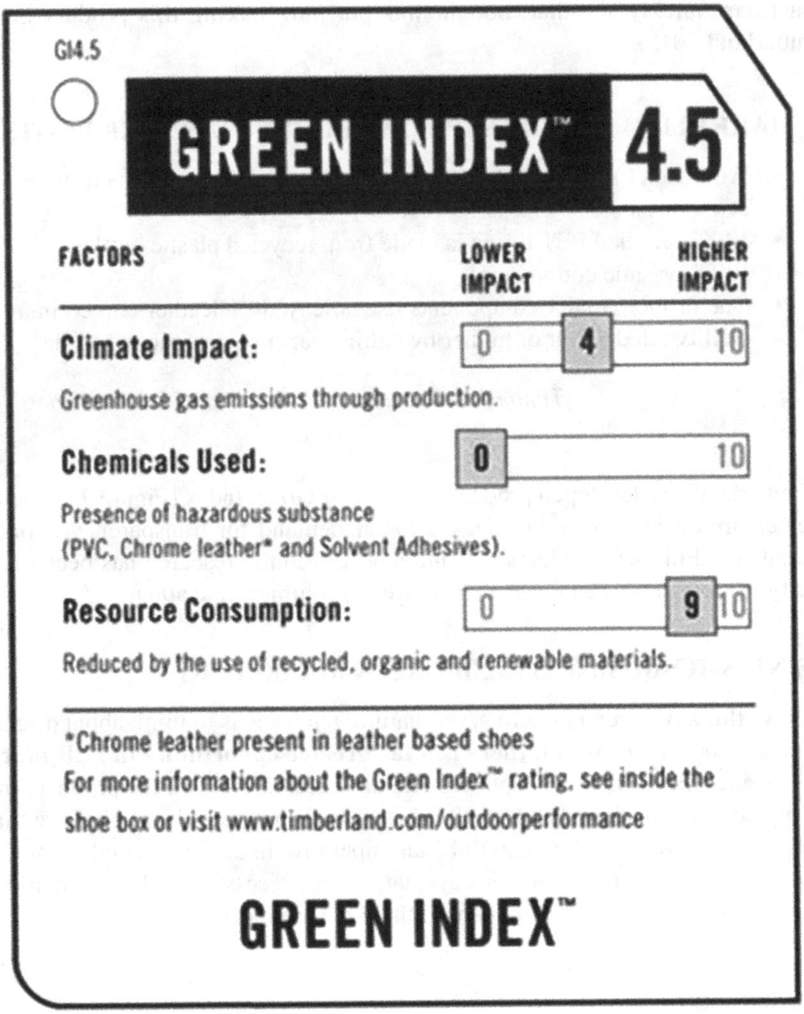

FIGURE 7.1 Timberline Green Index.

A program can't get any closer to meeting customer's needs than that. In addition, if you polled people throughout the globe on what is the most important environmental issue facing the world, inevitably you would hear global warming or climate change. GE's program sets to tackle this issue through this business program.

MEETING CUSTOMER DEMANDS WITH GREENER PRODUCTS

Ecomagination is a "groundbreaking strategy the company used to build more efficient machines that produce cleaner energy, reduce greenhouse gas emissions, clean water and cut its use, and make money while doing it" (GE 2015).

The program claims that there are many examples of greener products. An Ecomagination report boasts that they have some of the most efficient products in the world. Some examples include the Tier 4 Locomotive, which decreased emissions by approximately 70% or more over their Tier 3 product and saved customers an estimated $1.5 billion. The HA-turbine is the largest, most fuel-efficient gas turbine in the world, at more than 61% efficiency. The LEAP jet engine gives customers a 15% improvement in fuel efficiency versus its predecessor and provides improvements in noise and emissions, and the lowest overall cost-of-ownership in the industry—a critical aspect to the airline industry. The city of San Diego employed Ecomagination's "intelligent" lighting system, called LightGrid, which links its streetlights to the Industrial Internet. The city replaced more than 3,000 light fixtures with GE LED lights, their intelligent street lighting systems can reduce electricity consumption by 50–70%. If you were in the market for one of these products, surely you would take a good look at Ecomagination products. The key messaging is that **it saves money while providing significant environmental improvements**.

Ecomagination has been an extremely successful green marketing program. They have developed a robust portfolio of greener products (see the case study in the Developing Greener Products chapter). The whole point of the program is to meet customers' needs. Marketing is positioned and connected right at the core requirements—more energy- and water-efficient and clean-power generating products. They have revolutionized the way green marketing communication is conducted, and the case could be made that GE has brought B2B green marketing mainstream due to its success and market penetration.

BASF

Surprisingly, the chemical industry is one of the most beneficial categories to emphasize the greener aspects of products. I recall seeing television commercials years ago by BASF, a German chemical company, saying, "We don't make a lot of the products you buy; we make a lot of the products you buy better." Today, they can say, we make the products you buy "greener."

BASF's platform for communicating their more sustainable products is called Sustainable Solution Steering®. This process puts all products into four categories tied to providing sustainability solutions to their customers. The "accelerator" category is the highest level and is described as products that provide a "solution with a substantial sustainability contribution in the value chain." As a chemical company, they believe chemistry is an enabler offering "business opportunities" for meeting customer needs.

One of the customer groups that BASF services is automotive. Focusing on energy efficiency and reducing air pollution are key sustainability concerns for this industry. To address this need, BASF offers lightweight plastic materials which lower the weight of the cars, resulting in better fuel-efficiency and catalysts which reduce exhaust emissions to improve air quality.

Another category that they sell into is packaging. Here, the use of materials that are lighter weight or biodegradable are sustainability delighters for their customers. An accelerator product in this area would be the biodegradable eco-vio® paper coating, which enhances the proper disposal of paper-based products. Another would be the water-based resin Joncryl FLX®, which is an alternative to solvent-based printing technologies, reducing volatile organic compounds emissions from the printing process on packaging and the overall packaging life cycle (BASF 2014).

Chemical companies have a strong incentive to focus on the sustainability needs of their customers, since their products are the building block for all products. BASF is meeting the demands of customers through products that lower environmental impact such as lightweight plastics. Green claims are backed up by their internal sustainability rating system for all products, which also enables communication of the greener attributes to their customers.

STEELCASE

Every business needs office furniture, and most companies now have an environmentally preferred purchasing program that addresses office furniture and equipment. Sustainability concerns for office furnishing revolve around materials used and the end of life of the product. Businesses avoid products like flame retardants, toxic metals, and wood from endangered forests. When it's time to get new furniture, they don't want the old equipment to go into a landfill. Steelcase addresses these issues head on in their messaging for their products.

Steelcase is an over 100-year-old company headquartered in the United States that sells globally. Their product portfolio includes chairs, tables, bookcases, lighting, and screens—basically anything you need for your office to operate. The raw materials for these types of products include plastics, metals, wood, and leather, all of which have their own set of environmental issues associated with them. Recognizing the importance of purchasing more sustainable office furnishings to their customers, Steelcase has "customers" at the center of their sustainability focus. Their vision is listed as follows:

Our sustainability vision is clear: bring lasting value to our customers, employees, shareholders, partners, communities and the environment.

To address the use of materials and end-of-life issues they have deployed design for the environment and circular economy thinking. "We design products for circularity by avoiding and eliminating materials of concern, optimizing performance throughout the life cycle and for remaking, recovery, and end of life strategies. We are actively working with our supply chain to eliminate and phase out materials of concern and to develop suitable alternatives where they may not yet exist" (Steelcase 2023).

If you were in the market for new furniture, it would be great if the company you are buying is also responsible for managing your old items. In their 2021 sustainability report, Steelcase indicates that they redeployed 84,000 workstations, with an average of 32% of reuse, 68% of recycling, and recycling with 83% of raw material reuse (Steelcase 2023).

STEELCASE END OF USE MESSAGING

Our products are built to last; sometimes they outlast customer needs. Therefore, we offer multiple programs to extend a product's lifespan through reuse and recycling, refurbishing, and donating. These end-of-use services keep furniture out of landfills, provide non-profit organizations with needed resources, and help customers meet their sustainability goals (Steelcase 2023).

An excellent practice employed includes the use of product environmental profiles. These documents tell a customer everything they need to know about key sustainability elements. Evaluation of the environmental profile of the Siento® office chair addresses issues that are important to customers, such as certifications that give customers confidence regarding the absence of materials of concern and how it contributes to LEED certification. Below is part of the data that appears in the environmental profile.

SIENTO® OFFICE CHAIR ENVIRONMENTAL PROFILE INFORMATION

Certification: Cradle to Cradle Certified®—Silver (depending on options), SCS Indoor Advantage™ Gold certified for indoor air quality in North America, level® 3 certified to ANSI/BIFMA e3 standard.

LEED contribution: Recycled content, regional materials, rapidly renewable materials, low-emitting materials, sustainable purchasing, innovation in design.

Environmental facts: Ninety-one percent recyclable, 28% recycled content, life-cycle assessment completed, PVC-free and chrome-free, standard leather is chromium-free.

Reviewing this environmental profile gives customers confidence that sustainability issues have been well managed. It also makes it easy for customers to calculate the environmental improvements realized from their purchase by listing the percentage and total weight of recycled content and how they can get LEED points.

Steelcase is a great example of B2B green marketing. They make customer concerns the center of their marketing program. Additionally, there are many good methods deployed to entice customers to buy from their company. The use of third-party certifications, product environmental facts, providing data for the customer to easily calculate their environmental improvements through purchase of their products, and providing end-of-life solutions are all great ways to sell products to the growing number of companies that want to purchase more sustainable office furniture and equipment.

SODEXO

Sodexo, Inc., the largest **food service company** in the world, has also embraced sustainability through their ***Better Tomorrow Plan***. There are three core pillars to this plan:

We are: The fundamentals that are the cornerstone of a responsible company.
We do: Four priorities—a responsible employer, nutrition/health/wellness, local communities, and the environment.
We engage: Dialog and joint actions with our stakeholders.

(*Sodexo 2022*)

One of their commitments, to source sustainable fish and seafood, has resulted in a good green marketing story. It is reported that 7 of the top 10 marine fisheries are over-fished. Sodexo has pledged to make a "positive impact on the health of the world's oceans and fisheries by significantly reducing the amount of unsustainable seafood in the food service industry." To honor their commitments, they signed a global agreement in 2011 with the Marine Stewardship Council (MSC), a non-profit independent organization that certifies that wild-caught fish are not species which are at risk by overfishing. Sodexo also has partnered with World Wildlife Fund (WWF) and works with other nongovernmental organizations to be advised regarding **responsible seafood sourcing**. Traceability certification ensures its clients and consumers that MSC-certified products are not mixed or replaced at any stage of the supply chain with non-certified seafood (Sodexo 2022). Making a commitment like this will give Sodexo a competitive advantage for customers that are interested in purchasing a sustainable product.

Having certified seafood will make it easier for Sodexo to sell to customers that are being pressured by environmental groups because of the over-fishing that has occurred. Sodexo has developed a greener product to meet customer needs—a certified sustainable food. They have clearly committed to responsibly sourcing seafood and communicating their company's greener benefits. A partnership with an independent nongovernmental organization to certify their products sustainable benefits helps to solidify their products credentials and protects against green washing.

SEALED AIR DIVERSEY CARE

Diversey Care, a business unit of the Sealed Air company, manufactures and markets detergents, cleaners, sanitizers, lubricants, floor care products, carpet cleaners, and carpet cleaning and floor care machines, along with a host of other products and services. Their primary customers are hotels, hospitals, and companies that operate office and manufacturing buildings. While most companies discuss sustainability on their website as one of their objectives, Diversey has it integrated into their business model.

A key to green marketing is to know your customers' needs and to help them with their sustainability goals. This commitment is plainly stated when describing their products: "Diversey believes that the greatest opportunity for positive environmental and social impact does not lie solely in its own operations, but also through the delivery of products and services that enable our customers to minimize their environmental footprint and operate more efficiently" (Diversey 2023).

A perfect illustration of Diversey putting sustainability into practice is their Pur-Eco Chemical products. These industrial cleaning products all meet the EU Ecolabel and Nordic Ecolabel requirements and are claimed to be biodegradable and are formulated with raw materials derived from natural vegetable sources. They also sell their products in concentrate formulas, which result in less storage requirements and packaging waste for customers.

The messaging used focuses on business customers' concerns for effective products with low environmental impact that is free of toxic chemicals. Some of the key communications in their marketing materials include:

- Highly effective **"green" cleaning** for a safe and healthy work environment.
- Ultra **Concentrated Chemicals** for a "greener" future. Product concentration leads to more efficient use of chemicals and packaging material and less impact from transportation and storage.
- **Reduced** plastic waste by 65%, cardboard waste by 50%, and carbon dioxide emissions by 62%.
- **Products are biodegradable** and are formulated with raw materials derived from vegetable sources.
- **Independent certifications** of reduced environmental use of third-party eco-logos.

Diversey has built sustainability into the core of their product offerings. Getting third-party certifications to reinforce to their customers that they indeed have greener products is a smart practice. Their messaging covers all the most important things to their customers: safe, effective products with lower environmental impacts. The green story is convincing due to their greener product credentials, top management support, and the fact that sustainability is woven into the company's DNA.

Key Elements of Effective Green Marketing

We have seen that effective green marketing programs, whether B2C or B2B, have certain key elements:

1. Greener products are woven into the business strategy.
2. Understanding customers' desires and goals and align greener products to meet these needs.
3. Clearly communicate greener characteristics with third-party certifications or company-branded programs. Use of communication tools like environmental product profiles or company-generated labels.

4. Be authentic and credible in all marketing efforts, substantiate all claims, and be transparent.
5. Sustainable branding is an enhancement to other brand qualities—the idea that it's a great product "**and**" it has these sustainable attributes. A product's greener quality should never overshadow its purpose.

We have seen in our analysis of green marketing approaches that these elements have been successfully put into practice by consumer package goods firms, food suppliers, chemical manufacturers, electronics, and many others. The most successful green marketing programs include all the key elements. As stated in the beginning of this chapter, a brand must be built on the foundations of:

a. Having a credible greener product story.
b. Meeting customer demands.
c. Appropriately communicating the greener attributes.

REFERENCES

BASF. (2014). Sustainable Solution Steering. *ResearchGate*. https://doi.org/10.13140/2.1.3032.0649

Diversey. (2023). *Sustainability | Diversey*. Diversey Global. Retrieved August 27, 2023, from https://diversey.com/en/sustainability

EcoWatch. (2015, May 18). 10 Big announcements big business made to meet consumer demand for green products. *EcoWatch*. Retrieved August 27, 2023, from http://www.ecowatch.com/10-big-announcements-big-business-made-to-meet-consumer-demand-for-gre-1882042242.html

Fast Company. (2010). Suppliers Set Out to Grade Products with Sustainability Scorecards. *Fast Company*. https://www.fastcompany.com/1693714/suppliers-set-out-grade-products-sustainability-scorecards

GE. (2015). *Ecomagination Ten Years Later: Proving that Efficiency and Economics Go Hand-in-Hand | GE News*. Retrieved August 27, 2023, from https://www.ge.com/news/reports/ecomagination-ten-years-later-proving-efficiency-economics-go-hand-hand

GE. (2016). *GE 2015 Annual Report*. Retrieved August 27, 2023, from https://www.annualreports.com/HostedData/AnnualReportArchive/g/NYSE_GE_2015.pdf

Grundmann, G., Klein, F., & Josten, F. (2022). Sustainability in business: Staying ahead of the curve. *Deloitte Insights*. https://www2.deloitte.com/us/en/insights/topics/strategy/sustainability-in-business-staying-ahead-of-the-curve.html

Guevarra, L. (2010, May 4). *Kaiser applies new Green Scorecard to $1B medical supply chain | GreenBiz*. Retrieved August 27, 2023, from https://www.greenbiz.com/article/kaiser-applies-new-green-scorecard-1b-medical-supply-chain

Hollender, Jeffery, Ashley Orgain, and Ted Nunez. (2010). *Toward a Better, More Effective Brand of Green Marketing. Sustainability Solutions Paper.*

Honest Tea. (2020). *2020 Honest Mission Report*. Retrieved August 27, 2023, from https://rganic.com/content/dam/nagbrands/us/honesttea/en/mission/CCD20-231-HT-Mission-Brochure-R10.pdf

Lowe's. (2021). *2021 Corporate Responsibility Report*. Retrieved August 27, 2023, from https://corporate.lowes.com/sites/lowes-corp/files/CSR-reports/Lowes_2021_CSR_06.03.22.pdf

Mckinsey. (2023. *Consumers care about sustainability—and back it up with their wallets.* McKinsey & Company. Retrieved August 27, 2023, from https://www.mckinsey.com/industries/consumer-packaged-goods/our-insights/consumers-care-about-sustainability-and-back-it-up-with-their-wallets

NYU Stern. (2023). Effective Sustainability Communications. In *NYU Stern.* Retrieved August 27, 2023, from https://www.stern.nyu.edu/sites/default/files/2023-06/Jun%20 19%20FINAL%20-%20NYU%20Stern%20CSB%20and%20Edelman%20Impact%20 -%20Claims%20That%20Cut%20Through%20-%20A%20Guide%20For%20 Brands%20To%20Commercializing%20Sustainability%5B89%5D%5B75%5D.pdf

Organic Trade Association [OTA]. (2023, May). *U.S. Organic Industry Survey 2023.* Retrieved August 26, 2023, from https://ota.com/news/press-releases/22820

Seventh Generation. (2023). *Hand Wash – Free & clean.* Retrieved August 27, 2023, from https://www.seventhgeneration.com/hand-wash-free-clean-unscented

Shelton Group. Eco Pulse 2022. Eco Pulse 2022 Segmentation slides 12.20.pdf.

Sodexo. (2022). *2022 Sustainability and Corporate Social Responsibility Report.* Retrieved August 27, 2023, from https://us.sodexo.com/files/live/sites/com-us/files/inspired-thinking/2022/CSR-Report-2022.pdf

Steelcase. (2023). *Environmental.* Retrieved August 27, 2023, from https://www.steelcase.com/discover/steelcase/esg-overview/environmental/#highlights

Tesco. (2021, September 24). Tesco commits to net zero emissions from its supply chain and products by 2050. *Tesco.* Retrieved August 27, 2023, from https://www.tescoplc.com/tesco-commits-to-net-zero-supply-chain-and-products-by-2050/

The Clorox Company. (2010). *2010 Clorox Corporate Responsibility Report.* Scribd. https://www.scribd.com/document/39139385/2010-Clorox-Corporate-Responsibility-Report

Timberland. (2015). *Timberland Green Index.* Retrieved August 27, 2023, from http://web.archive.org/web/20190624091722/http://greenindex.timberland.com/

Timberland. (n.d.). *Men's Timberland® Originals 6-Inch Boot.* Retrieved August 27, 2023, from https://www.timberland.com/en-us/p/men/footwear-10039/mens-timberland-originals-6-inch-boot-TB015551210

Unilever. (2023). Sustainable and regenerative sourcing. *Unilever.* https://www.unilever.com/planet-and-society/protect-and-regenerate-nature/sustainable-and-regenerative-sourcing/

Werbach, A. (2009). Harvard Business Publishing Education. In *Strategy for Sustainability: A Business Manifesto* (pp. 109–110). https://hbsp.harvard.edu/product/3424BC-PDF-ENG

Williams, E. (2015). *Green Giants: How Smart Companies Turn Sustainability into Billion-Dollar Businesses.* AMACOM.

8 Consumer Interest in Sustainable Purchases Continues to Grow

Suzanne Shelton

INTRODUCTION TO THE CHAPTER

I am privileged to be able to share market research and advice from a leading sustainability advertising agency in this book. Shelton Group, an ERM Group company, is a thought leader in the green marketing space and has an excellent feel for the demand for greener products in the United States.

As we will see in the information presented in this chapter, the pull for greener products is growing and is predicted to continue. In fact, Shelton's 2022 Eco Pulse™ market research indicates that 75% of US consumers say they are searching for greener products.

Before initiating a green marketing program, it is critical to understand your consumer. Shelton Group has segmented consumers into four groups based on their attitudes and behaviors related to sustainability (Actives, Seekers, Indifferents, and Skeptics). That segmentation can help a brand team identify the best way to position their product toward their target consumers. The concepts presented here are the backbone to approaches and methods presented in the subsequent chapters on how to effectively market a greener product.

As we will see in the evaluation of the most successful green marketing programs, they have implemented the advice in this chapter and truly "get it" when it comes to the proper way to position a product. Some groups may be willing to pay a premium for greener products (Shelton calls them the Actives), while others want greener products but want them as an "and." What I mean by this is that they want a product to be effective, at their price point, "and" if it doesn't cost more, the greener attribute will push the customer over the edge to buy it. Knowing how to position a brand to its target group and understanding which certifications or attributes (e.g., certified organic) are meaningful and can make you a winner. Understanding message preferences is just as important as green product attributes, **because what good is a greener product if you can't get the customer to buy it?**

The data in this chapter are based on surveys of the US population as well as 11 other countries. Of course, there are regional differences, but understanding your consumer and knowing the right questions to ask is applicable everywhere.

DOI: 10.1201/9781003441939-11

GREEN PURCHASE DRIVERS DIFFER BY COUNTRY

Individuals look at the world and companies differently. Every distinctive perspective is shaped by their culture and country. Shelton Group's annual Eco Pulse™ went global in 2023 and surveyed people in 12 different countries to understand how they envision an eco-friendly person and the impact that brings to a company's sustainability communications (Figure 8.1). This online survey was fielded between February and March 2023 and had a total of 5,497 respondents divided into the Americas, Europe, Middle East/Africa, and Asia-Pacific (APAC). This study used Hofstede's Cultural Dimensions Theory as a framework for comprehending intercultural dynamics in both business and academic contexts. Hofstede's Six Cultural Dimensions include the *Power Distance Index, Individualism vs. Femininity, Uncertainty Avoidance index, Long-Term orientation vs. Short-Term Normative Orientation*, and *Indulgence vs. Restraint.*

POWER DISTANCE INDEX AND INDIVIDUALISM VS. COLLECTIVISM

These two dimensions were found to be associated with a country's sustainability performance and path toward sustainable development. The Power Distance Index measures how power is distributed among individuals. It analyzes to what extent people expect, accept, or question authority. While Individualism vs. Collectivism measures self-image defined in terms of "I" or "we," it also measures the scope of responsibility individuals feel toward other people. Individualist societies

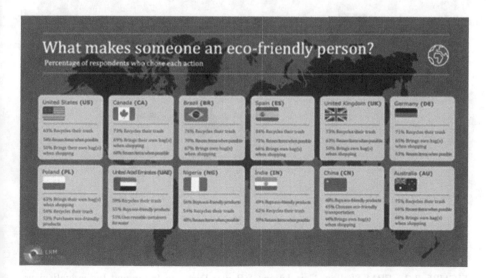

FIGURE 8.1 ERM Shelton, in-depth global survey breakdown. (From Cultures, Countries & Your Sustainability Story: Eco Pulse™ 2023, Shelton Group.)

tend to encourage people to look after themselves and their families, while collectivist societies encourage people to belong to "in groups" where mutual care is exchanged for loyalty. Individualism is almost always associated with Low Power Distance, while collectivism is almost always associated with High Power Distance (Figure 8.2).

In the 7 out of 12 countries that were individualist, Shelton Group found that inflation is a greater concern for people than key environmental or social issues. In the five **Collectivist countries, people were more concerned about visible environmental damage,** including air and water pollution, as well as issues related to food and water shortages. **Purchasing eco-friendly products was also perceived as a top trait of an eco-friendly person according to 5 out of the 12 countries (4 being Collectivist).** Now, looking at the Power Distance Index more closely, countries with a Low Power Index had similar concerns about inflation to individualist countries (five out of the seven most concerned with inflation had Low Power Distance). Recycling was identified as the primary behavior associated with an eco-friendly person, with 8 out of 12 countries mentioning

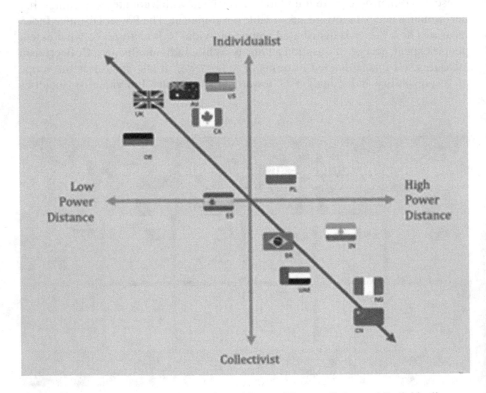

FIGURE 8.2 ERM Shelton, Correlation of Power Distance Index and Individualism vs. Collectivism. (From Cultures, Countries & Your Sustainability Story: Eco Pulse™ 2023, Shelton Group.)

recycling. Notably, five out of those eight countries are characterized by Low Power Distance.

Globally, recycling is considered the bare minimum that can be done for the environment, and it is considered a top defining characteristic of an eco-friendly person in most Individualist/Low Power Distance countries. This is simply because recycling and reusing is an individual behavior and gives consumers control over one's personal impact. However, people in Individualist/Low Power Distance countries lack confidence in the recycling system. Some marketing applications that can be applied to these countries would be to target your messages, counteract skepticism with transparency, educate on the recycling process, and promote recycling success stories. For example, explain how your company designs its product and packaging for easy recallability and how you're helping improve the recycling processes. Use case studies, visuals, and infographics to promote education and help consumers take back their power to make a difference.

In High Power **Distance/Collectivist countries** (Poland being an outlier) consumers consider **purchasing eco-friendly products as a top trait in an eco-friendly person.** Purchasing eco-friendly products requires personal motivation and can be satisfying for people who feel personally responsible for climate change. This behavior, unlike recycling, does not require participation in a system or formal structure, this gives back the power to the individual. This also plays a big role in how individuals shape their public identity. Most people in a High Power Distance/Collectivist country want to be seen as someone who purchases eco-friendly products (Figure 8.3). Some marketing strategies would be to lead with a sustainable lifestyle, emphasize the benefits of eco-friendly products, and ensure quality and performance assurance. Some examples would be to use your brand image to contribute to forming your consumer's personal image, showcase eco-friendly products and their features/benefits, and present customer reviews that confirm the effectiveness of your products.

The topic that might come as a surprise is that all individualist countries were concerned about inflation. Most people even said they look for more energy-efficient, natural, eco-friendly products. Some marketing strategies for **Low Power Distance/ Individualist** countries (and Poland) could be to **use economic messaging**, highlight the immediate economic benefits, offer lifetime cost analysis, and source locally for job creation. Some examples are to market the financial benefits such as energy-efficient appliances leading to lower utility bills and how energy-efficiency can help mitigate utility rates increasing in the future.

Lastly, almost all Collectivist/High Power Distance countries (not including Poland) are concerned with visible environmental damage. Many of these countries' concerns are directly related to climate change, such as deforestation or water shortages, which confirms that they not only believe in climate change but also that it is correlated to human activity. This leads to the unsurprising conclusion that these countries want to hear from companies about their greenhouse gas emissions (GHG) initiatives. **A majority of people even say their opinion of a company would improve if the company were a leader in reducing its GHG emissions**. Some marketing strategies would be to begin with tangible benefits,

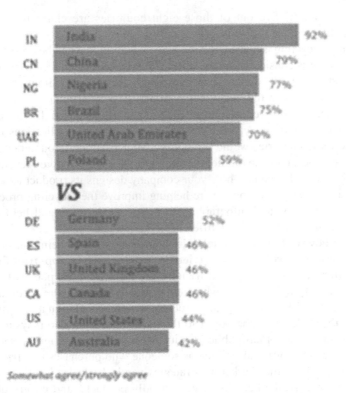

IN	India	92%
CN	China	79%
NG	Nigeria	77%
BR	Brazil	75%
UAE	United Arab Emirates	70%
PL	Poland	59%

VS

DE	Germany	52%
ES	Spain	46%
UK	United Kingdom	46%
CA	Canada	46%
US	United States	44%
AU	Australia	42%

Somewhat agree/strongly agree

FIGURE 8.3 ERM Shelton, "How much do you agree or disagree with the following statement? *Buying/using eco-friendly products is an important part of my personal image.*" (From Cultures, Countries & Your Sustainability Story: Eco Pulse™ 2023, Shelton Group.)

pursue visual impacts, communicate about your GHG emissions and initiatives, and even partner with environmental organizations. Some examples include, using before and after imagery to showcase how your products can make a difference, speaking about operational and product climate impact, and collaborating with organizations to show your brand's commitment. Many **collectivist countries agree** that volunteering/**giving to an environmental cause is a top eco-friendly behavior** as well, so communicate about how your brands actively work in communities to improve an aspect of the environment or society that makes sense for your brand.

GREEN PURCHASE DRIVERS DIFFER BY LIFE STAGE

Shelton Group has also created a report that comes from their Good Company survey data, which was fielded from December 2020 to January 2021 and includes 2,000 respondents from America. This survey explores how and why Americans view companies as "good." The finding revealed that an individual's life stage has a big impact on their buying behavior. In the first part of adulthood, age, income, and spending increase simultaneously. In the second half of adulthood, age, income, and spending decrease simultaneously.

Every year, Shelton Group asks Americans: "Have you ever chosen a product or service over another or stopped purchasing a product or service based on the social or environmental record of the brand?" Thirty-three percent answered "yes," and 26% of Americans can name the exact brand, unaided.

The mainstreaming among Americans believing that climate change is occurring, due to human activity (57%) has led to the integration of green product purchases being tied to their identity (42%). Every year, Shelton Group asks Americans: "Have you ever chosen a product or service over another or stopped purchasing a product or service based on the social or environmental record of the brand?" Thirty-three percent answered "yes," and 26% of Americans can name the exact brand, unaided. This led Shelton Group to divide the patterns of "Shoptivism" into three categories: *choosers*, *stoppers*, and *changelings*.

Choosers have been identified as individuals who choose a product or service due to the brand's social or environmental record. **Stoppers** have decided to stop the purchasing of a product due to the brand's social or environmental record. **Changelings** have both chosen a product and stopped using a product due to the brand's social or environmental record. Although these groups have many differences, they do have similarities in demographics. All three groups are equally likely to be any gender, Democrats, from all income ranges, any marital status, and "Actives" (elaborated on later in the chapter). All the groups are highly likely to believe climate change is occurring and caused by human activity and all are highly likely to feel personally responsible for changing their daily habits and purchasing practices to do something positive for the environment.

The differences begin with Choosers being more likely to be millennials, spend 6+ hours on social media every week, are educated (especially on sustainability-related issues, 35%), homeowners, and city-dwellers. Younger people, especially those with young kids, purchase more things and end up picking brands that will become part of their lives. Sixty percent of **Choosers** express trust in the companies they purchase from to be environmentally/socially responsible. This group of consumers is influenced by marketing and messaging, and 45% trust larger companies more than smaller ones. Messaging to this group can begin with providing them with your brand's stand on social or environmental issues via social media platforms. Testing shows that **a brand can increase its favorability dramatically with this group by communicating a strong social or environmental stance**.

Stoppers are more likely to be boomers, retired, and living out of the city. **Stoppers** are at the point in their lives where they are downsizing and dropping brands from their lives. This group is also more likely to be apathetic toward companies and their sustainability issues. However, once they learn of a brand's social or environmental efforts, there is an uptick in favorability. **Messaging to this group can begin with emphasizing employee treatment, as they are the most passionate about this out of the three groups**. It is also important to show your brand as

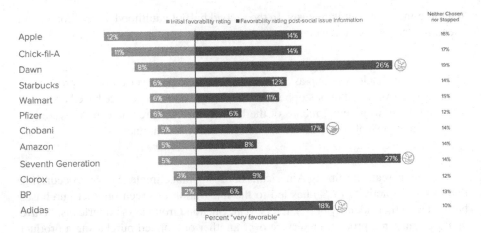

FIGURE 8.4 Changelings uptick in favorability of a company after their message is learned. (From Shoptivism, Eco Pulse™ 2022, Shelton Group.)

community-oriented, such as making charitable donations. Finally, environmental messaging can be impactful but only after the people aspect. Good old-fashioned television advertising is the best way to reach this group.

Changelings are the final group, and they are most likely to be Zoomers (Gen Z). This group usually cares more about the environment, even over their own comfort and convenience. **Sixty-five percent of Changelings do not trust large companies as much as they do smaller companies**, and 55% are concerned that **if a company does not communicate about its environmental or social initiatives, then they are not trustworthy**. Changelings are also much more likely to see a brand positively once sustainability messaging and stands are provided (Figure 8.4). Currently, the best way to communicate a brand's social or environmental story to Changelings is through social media (specifically TikTok).

THE MAINSTREAMING OF SUSTAINABILITY

In 2015 the first Buzz on Buzzwords study was conducted, and the findings revealed that green was a selling point for major global brands. This study helped analyze what consumers thought of the **Big Three umbrella terms**, *green*, *eco-friendly*, and *sustainable*. The updated 2022 survey of 2,000 American consumers showed that **attitudes toward environmental topics have changed substantially**. This was done by asking a series of "slider bar" questions that determined how drastically the attitudes and beliefs toward the Big Three have changed. The beliefs and attitudes that were considered toward the word "green" were:

- Positive or negative
- Easy or difficult to understand
- Good or bad for health
- Conservative or liberal
- Expensive or inexpensive

- Trustworthy or untrustworthy
- High quality or low quality
- Meaningful or meaningless
- Indicative of a good or bad brand/company

The reactions to the terms, "eco-friendly" and "sustainable," were also recorded and compared. In 2015 "green" was closely affiliated with politics. In the 1980s, "eco-friendly" was introduced and had a more positive connotation. The word "sustainable" has been used in early environmental movements, especially agriculture, but in the 2000s it became a business buzzword.

A shift in buzzword trends between 2015 and 2022 is the decline in the percentage of individuals associating the term "green" with an expensive product, dropping from the initial 76% to 42%. Additional shifts involve its political connotations. In 2015, 39% of Democrats perceived it as liberal, while 13% of Republicans perceived it as conservative. Jump to 2022 and now 51% of Democrats perceived it as liberal and 28% of Republicans perceived it as conservative. Another observation is that consumers understand buzzwords such as "green," "sustainable," and "eco-friendly," meaning that they are educating themselves on current issues, resulting in these terms becoming even more mainstream.

> A shift in buzzword trends between 2015 and 2022 is the decline in the percentage of individuals associating the term "green" with an expensive product, dropping from the initial 76% to 42%.

The key takeaway regarding the **utilization of buzzwords** in your marketing strategies is that **they are no longer linked with the notion of being expensive or as politically divisive as they were in the past**. However, because they are now mainstream, they can lose their impact. These words can also be seen as less impactful because consumers expect today's brands to be "green," "sustainable," and "eco-friendly," and **they do not believe they should pay more for that label.**

SO, WHO'S BUYING GREEN?

Shelton Group has developed a proprietary segmentation system to better understand the attitudes, behaviors, drivers, motivations, and personalities of those Americans who are buying green. The following data come from their Eco Pulse™ 2022 and 2023 studies. Four distinct consumer groups have been identified at different points along the green spectrum. The top two segments are the Actives and the Skeptics, followed by Seekers and Indifferents (Figure 8.5).

MEET THE ACTIVES – *34% OF AMERICANS*

The most involved of these groups is the Actives, representing 34% of the overall American adult population. This group is **green in both their beliefs and activities, with 86% reporting that they are actively searching for greener products.**

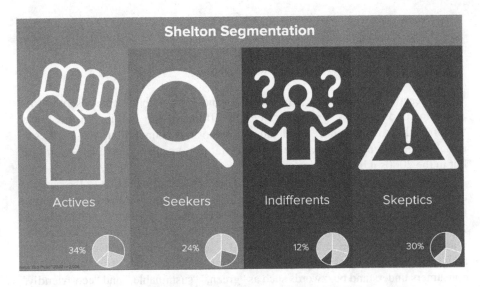

FIGURE 8.5 Shelton segmentation. (From Eco Pulse™ 2023 Shelton Group.)

Demographics

The Actives are well educated and have the disposable income to follow through on their green values:

- They're equally likely to be male or female.
- They're slightly older; 35–44 (skewing slightly older than March 2020).
- They're well educated, with significantly more graduate degrees.
- They're more likely to have a higher household income of $100,000+.
- They're more likely to be employed full time.
- They're more likely to have kids in the household.
- Significantly more likely to be Democrats.

Lifestyle, Green Attitudes, and Behaviors

Actives prioritize the environment over comfort or convenience, they feel a deep sense of personal responsibility to make an impact by changing their daily habits and purchasing choices. Their commitment extends to incorporating eco-friendly products as a significant part of their image. They believe climate change is occurring, and human activity is the primary cause. Their concerns encompass climate change, plastic in the ocean, gun violence, air pollution, and water pollution. However, in households with children, priorities shift to concerns about gun violence, plastic in the ocean, water pollution, climate change, animal habitat destruction, and air pollution. Overall, the presence of children in the household increases the level of concern across all environmental issues.

Actives are also significantly more likely than other groups to make their own cleaning products and grow some or most of their own food. They cook at home rather than going out to eat significantly more often than all other groups, and they are also more likely to be do-it-yourselfers around the house. These facts all point to an important personality characteristic of Actives: They are pragmatic "doers." They are more physically active than the overall population and are significantly more involved than other groups in sports – particularly outdoor activities such as hiking, kayaking, and camping – and the arts. Actives score very high on both green attitudes and behaviors. **This group prioritizes the environment higher than other groups, and that belief guides their actions**. To illustrate,

- Seventy-seven percent are searching for greener products.
- An average of 14 green behaviors or purchases were reported, compared to 11 overall.
- They will most likely stop purchasing and supporting a manufacturer based on their environmental record, and they will communicate it to others.
- They are most likely to state that a company's involvement in social issues or its nonprofit partnerships and donations has a strong impact on their own decision to buy its products.

This level of environmental engagement and concern translates into their definition of "green." To this group, green means several things: environmentally friendly, recycled or recyclable, and energy efficient – but it's also likely to mean water-conserving, safe, healthy, organic, and natural.

There's not much controversy about global warming with this group – three out of four of them believe it's real, happening, and caused by human activity. They feel a strong sense of personal responsibility to change their daily habits and purchase behaviors to positively impact the environment.

While they're searching for green products at a high rate, they come in second after Seekers who are at 88% (described below). This is partly because many **Actives would rather "do green" than "buy green"** – they've already adopted many sustainable behaviors, such as growing their own food or making their own cleaning supplies, which makes them somewhat less likely to seek green products in those categories. In short, we suspect some reject consumerism as a matter of principle. Actives may also be less likely to seek green products because they've already established green purchase behaviors and brand loyalties. But Actives are very open to trying new things; so as new green options emerge in new categories; they'll likely be early adopters.

When considering which products to purchase, Actives are the group most likely to have chosen one product over another or stopped purchasing a product based on the environmental record and practices of its manufacturer. Actives know a product is green because they do their research. They read about the ingredients, content, and energy savings on the package and will research it on the Internet. They also trust third-party certification and third-party endorsement when determining if a product is green (Figure 8.6).

MEET THE SEEKERS – *24% OF AMERICANS*

Seekers are the third-largest segment and are very similar to (and sometimes even stronger than) Actives in their environmental beliefs and attitudes. **All of them are**

Segment Comparison 2020 to 2022: A shift from Skeptics to Actives

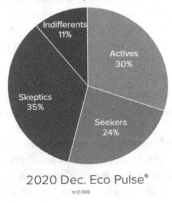

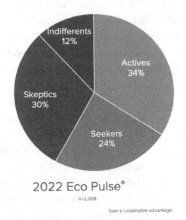

2020 Dec. Eco Pulse®
n=2,008

2022 Eco Pulse®
n=2,008

Gain a sustainable advantage.

FIGURE 8.6   Green segment percentage comparisons of 2020 and 2022. (From Eco Pulse™ 2023 Shelton Group.)

seeking **to be green**, but most fall somewhat short with their activities and purchase behaviors. With a higher average income than Skeptics and Indifferents, Seekers often do have the resources to act on their beliefs by making more expensive green purchases.

Demographics
Seekers are:

- Equally likely to be male or female.
- Slightly younger than Actives, predominantly Gen Z to Gen X.
- Not well educated; half have attained only a high school diploma or less.
- More likely to have a household income of $50,000 or less.
- White-collar or blue-collar workers.
- More likely than the overall sample to be minorities; they are the group with the highest percentage of Hispanics (19%).
- This group contains the highest percentage of non-homeowners (50% vs. 38% overall).

Lifestyle, Green Attitudes, and Behaviors
Seekers are less outspoken and spontaneous than Actives. They generally gravitate to the center of most personality scales. They mirror the overall population in terms of optimism, leadership, organization, and security. However, in our survey, they were more likely than average to say they give advice rather than ask for advice, indicating they are influencers. They are also somewhat more ambitious than Actives, but their reputation is more important to them than achieving a higher social status. Politically, Seekers are more likely to be evenly distributed based

on political leanings compared to previous years when Seekers tend to skew more Democratic. Seekers are primarily defined by their active search for greener products. However, **only 29% feel a strong personal responsibility to change their daily habits, including consumer practices**, to positively impact the environment. For instance,

- Seekers are moderately concerned about environmental issues, compared to the extremely concerned Actives.
- They believe climate change is primarily caused by human activity.
- Their comfort (49%) and convenience (28%) outweigh their concern for the environment, which is at 24%.
- Seekers are the segment most likely to search for greener products at 88%. However, only 51% view buying eco-friendly products as important for their personal image.

Meet the Indifferents – *12% of Americans*

Indifferents compose the smallest of the segments. This group has *decreased* since the 16% composed in 2010. However, during the 2020 to 2022 shift, it has *increased* by 1%. **This group is not highly impacted by a company's involvement in social issues, and they do not believe it is important for companies to take a stand on social issues**. Indifferents are also apathetic toward a company's efforts in social issues.

Demographics

Indifferents are:

- Slightly more likely to be female than male.
- They are more likely to be 55+ and primarily Boomers.
- They are more likely to be Caucasian.
- Significantly more likely to be retired.
- Eight percent are homemakers and 16% are unemployed, the highest of the categories.
- Usually much less educated.
- Low to moderate household income that ranges from $25,000 to 74,999.
- House size of 1–2 people and less likely to have children.
- Higher chance of being Republican.

Lifestyle, Green Attitudes, and Behaviors

Indifferents are not highly concerned about climate change, and they do not take much personal action. However, they do agree that climate change is happening and is caused by human activity, but do not feel responsible enough to change their habits to fight climate change. They are less likely to say that buying/using eco-friendly products is an important part of their personal image, and they care more about convenience than other segments (37%) but also care strongly about comfort (57%).

Only 24% say they are searching for greener products. When it comes to supporting companies, Indifferents are not very influenced by a company's involvement in social issues, or its nonprofit partnerships and donations.

MEET THE SKEPTICS – *30% OF AMERICANS*

Skeptics compose the second-largest segment and has *increased* from 26% in 2020. This group **does not believe it is important for a company to support social issues, and they are less likely to purchase a product based on the brand's support of social issues**. This would also mean that a company's involvement in social issues does not impact their decision-making of a brand or product.

Demographics
Skeptics are:

- More likely to be female (54%) than male (45%).
- Fifty-four percent are Gen X or older, which is lower than the 63% in 2020.
- Household is less likely to have children.
- Predominantly employed full time or retired.
- Eight percent are homemakers and 16% are unemployed, the highest of the categories.
- Almost half did not attend college.
- Half earn less than $50,000 and are less likely to be homeowners.
- Higher chance of being Republican, this has remained consistent.

Lifestyle, Green Attitudes, and Behaviors
Skeptics do not see "green" as a priority, and they care more about their comfort (61%) and their convenience (24%). Only 16% say they are searching for greener products, and buying eco-friendly products is not part of their image. That being said, it is not surprising that Skeptics participate in the lowest number of green behaviors and activities, an average of 6 compared to 11 overall.

SUMMARY AND CONCLUSION

In conclusion, the findings presented in this chapter offer a diverse amount of insights into the evolving landscape of green marketing and consumer perceptions and behavior. As environmental awareness becomes more integral in consumers' purchasing decisions, brands must navigate the preferences of their messaging as well as the delivery methods. Shelton Group's thorough and consistent analysis of segmentation through Actives, Seekers, Indifferents, and Skeptics provides the foundation for understanding green purchasing trends not only in America but globally. With the insight provided by the Eco Pulse™ 2023 global study, companies and brands can create targeted communications strategies that resonate with consumers from different cultural backgrounds and life stages, promoting an ever-growing sustainable, green, and eco-friendly marketplace for products.

ABOUT SHELTON GROUP, AN ERM COMPANY

Shelton Group is the nation's leading marketing communications agency focused in sustainability and corporate social responsibility. The company delivers thought-leading research, results-oriented marketing campaigns, emotionally compelling stories, and user-friendly designs – all intended to help companies gain a market advantage as they work to create a more sustainable future.

REFERENCES

ERM Shelton Group. (2023). *Eco Pulse™ Annual Report*. Cultures, Countries & Your Sustainability Story. ERM Shelton. https://info.sheltongrp.com/hubfs/Reports/Cultures-Countries-And-Your-Sustainability-Story-SheltonGroup-EcoPulse2023.pdf?utm_campaign=Engaging%20Middle%20America%20in%20Recycling%20Solutions&utm_medium=email&_hsmi=281902882&_hsenc=p2ANqtz-8vYd-SjaRGNqLz3qpDh3hGPZK6DoUVUN-CQOxdEmsZQuMXGeBd1iGiyZ7xLJTn-sUswEAxk01BRhFatn2PNQKNcV7eZ4mreZNNgHB_q7tqECaSlGHOU&utm_content=281902882&utm_source=hs_automation (Accessed December 18, 2023).
Shelton Group. (2022). Shoptivism: who are the people opting in and out of certain brands and why? https://sheltongrp.com/shoptivism-people-opting-in-and-out-of-certain-brands-why/. (Accessed April 26, 2024).
Shelton Group. (2023). *Eco Pulse™ Annual Report*. Eco Pulse Special Report. Shelton Grp. https://6711429.fs1.hubspotusercontent-na1.net/hubfs/6711429/Buzz-On-Buzzwords-Report-2022.pdf?utm_campaign=Buzz%20On%20Buzzwords&utm_medium=email&_hsmi=245323090&_hsenc=p2ANqtz-_y2qg-HVbWnc2iovJXOHlWWhrHd2q1gB1LAuXaRVTCswwW_90-JxX5IIU9BYTL_hSdfqG9QNYjDhe33kKHzY-uQIFF2N4lX-dqe9ClQinycYA5J-M4&utm_content=245323090&utm_source=hs_automation (Accessed December 18, 2023).

9 Marketing Green
Best Practices from OgilvyEarth

John Jowers and Ivellisse Morales

In the current marketing landscape, the ability to navigate fragmented media channels, infinite information, and dizzying distraction is crucial. Moreover, marketing greener products requires adding a sustainability lens. Ogilvy Earth, a marketing and communication firm, uses principles that can help marketers get their sustainability message across.

In recent years, the pursuit of brand purpose has caused an influx of sustainability commitments. The undeniable impact of climate change on humans, geopolitical turmoil, and mass migration has elevated sustainability as a strategic business driver.

In this chapter, we share our trusted best practices for green marketing, including:

1. Understanding the audience we are trying to reach so we understand where they are coming from.
2. Learning the triggers that drive the sustainable behaviors that we want to see.
3. Navigating the "Green Gap" and finding ways to eliminate the barriers to positive green behaviors.
4. Taking a leap of courage to become a purposeful brand that stands out in the marketplace.

THE NEW MARKETING BRANDSCAPE

It was a different world when David Ogilvy started Ogilvy & Mather in 1948 (Figure 9.1). World War II had ended. The soldiers had returned home. Manufacturing went from churning out war equipment to producing goods that made life a little bit easier. Brands were finding new ways to meet customer's needs, and the American people responded enthusiastically. The economy was stronger than ever.

Back then, advertising was art and copy. But in the decades that followed, the world changed fast, and the world of marketing changed with it. Technology transformed people's lives at an unprecedented pace. Brands turned to television to reach audiences live, and in color. Families grew fast as the "baby boom" took hold. Millions moved out of the cities to buy a better life in the suburbs. European immigrants began to be outnumbered by those from Latin America and Asia. The G.I. Bill gave birth to the emergence of a middle class. The American Dream became

 DOI: 10.1201/9781003441939-12

FIGURE 9.1 Photograph © Ogilvy & Mather, all rights reserved.

more attainable to more people. All the while, the beliefs, attitudes, and behaviors of the American public became much less easily defined, making audience appeals an increasingly complex task.

In the current marketing landscape, the ability to navigate fragmented media channels, infinite information, and dizzying distraction is crucial. There are principles from the David Ogilvy era that still ring true in today's digital world. Influenced by his research work at Gallup, Ogilvy emphasized the **disciplined study of the consumer**. In fact, Ogilvy was known to relentlessly pursue insights that often became the focus of campaigns.

One of our favorite Ogilvy-isms is unequivocally simple: **fail to understand the people you're selling to, and you'll fail to understand their needs**.

In today's world, a brand cannot only be useful, it has to be purposeful or it quickly moves towards becoming irrelevant.

Today, consumers expect brands to go beyond functionality. Brands are expected to demonstrate compassion, to be accountable, to be trustworthy, to offer a hand to help those in need. Brands must show the type of human qualities we expect from those we invite into our homes. A brand must be purposeful, or else it quickly moves toward irrelevance.

The pursuit of brand purpose has caused an influx of sustainability commitments. The undeniable impact of climate change on human, geopolitical turmoil, and mass migration has elevated sustainability as a strategic business driver. One area where sustainability strategy can falter is truly understanding consumers and effectively communicating with them to result positive yet sustainable business growth.

This chapter outlines an approach to crafting a well-positioned sustainability story. As marketers, we already know part of the answer: carefully cater to consumers. However, to market green products effectively, brands must mature from simply *existing to sell* to actually *doing good by doing well*.

None of this is revolutionary, nor even particularly novel to most green marketers. There's no shortage of sustainability communications experts who will offer their secret marketing sauce or proprietary research. Our aim is not to bombard you with data points, trends, or information to justify marketing green products. Instead, we want to apply a practical sustainability lens to understand what motivates consumers to be more sustainable.

In this chapter, we share our trusted best practices for green marketing, including:

1. *Understanding the audience* we are trying to reach, so we understand where they are coming from.
2. *Learning the triggers* that drive the type of sustainable behaviors that we want to see.
3. *Navigating the "Green Gap"* and finding ways to eliminate the barriers to positive green behaviors.
4. *Taking a leap of courage* if you want to become a purposeful brand that stands out in the marketplace.

> At OgilvyEarth—Ogilvy's sustainability communications team—we put David Ogilvy's principles into practice to help bridge the gap between complex sustainability issues, brand purpose, and consumer needs through strategic, integrated communications.

PUTTING PEOPLE FIRST

In Ogilvy's white paper *Mainstream Green*, we explored the "Green Gap"—the glaring disconnect between attitude and action as it relates to more sustainable behaviors, practices, and purchases. In our research, 80% of Americans ranked green activities, such as buying local food and recycling, of high importance, but only 50% confirmed they actually do these activities (Bennett and Williams 2011). Marketing can play a leading role in solving this classic conundrum, but we first need to understand our customers.

Start with Humans

Who are they? Who and what are they influenced by? What do they care about? By tapping into human behaviors, identities, experiences, values, anxieties, and joys, you can send the right and resonating message. You have to start with humans.

Human-centered design has risen in popularity, bringing together business strategy, innovation, and behavior science to uncover inspiring insights that lead to impactful solutions. Consumer needs can unlock answers to virtually any problem.

Empathy exercises such as immersion and in-depth interviews contextualize customers, enabling a deeper look into lifestyles, habits, beliefs, and values. These insights help identify the exact needs that will motivate and inspire consumers toward desired actions or behaviors. "Doing your homework" sets the context for the problem at hand.

In 1943, psychologist Abraham Maslow developed a five-tier model of human needs, ranked in hierarchical order (see Figure 9.2). His theory states that human deficiencies, or unmet needs and desires, subconsciously motivate us to act (Maslow 1943). The first tier—physiological needs—captures our most basic needs for survival. We need to feel safe and secure and have our basic needs met before we can focus on growing our interpersonal and social relationships, developing our self-esteem, and reaching self-actualization.

As green marketers, we spend a disproportionate amount of energy and dollars trying to change people's beliefs, values, and attitudes toward sustainability. **Our assumption is that the motivation, receptivity, and willingness to adopt green purchasing behavior depend on the customer's current needs and context.**

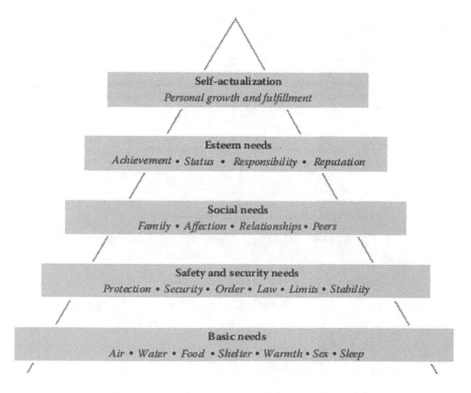

FIGURE 9.2 Maslow's hierarchy of needs (1943).

Inspired by Maslow's motivational theory, we developed the *Ogilvy Hierarchy of Green* as an interpretative, human-centered framework for understanding green behaviors and needs. Determining your customer's current tier will help you position and communicate sustainability messaging more effectively.

Ogilvy's Hierarchy of Green will help activate existing green impulses hidden within your customer (Figure 9.3). We must first identify needs, which motivate behaviors, which shape beliefs and attitudes. Our research reinforces what neuroscience research and behavioral economics have shown: attitudes and beliefs are shaped by behaviors, not the other way around.

Let's explore each tier of the hierarchy:

- *Basic Needs*: We all need food to eat, water to drink, a place to live, and other necessities to thrive in our day-to-day lives. It's hard to think about polar bears in the North Pole if you are trying to make ends meet or have serious health issues.[1] Immediate basic needs also trump any aspirational motivation to move upward in the hierarchy.
- *Safety & Stability Needs*: After our basic needs are met, we seek to establish stability, which lends to a sense of security. Stable employment and retirement savings offer financial security. Health insurance,

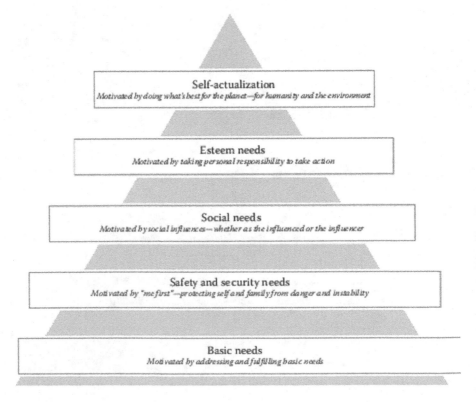

FIGURE 9.3   Ogilvy hierarchy of green.

for example, provides protection and financial support. Purchasing decisions are driven by cost, convenience, and impact on personal and family health.

- *Social Needs*: Humans are conforming creatures. Whether we realize it or not, we are heavily influenced by everyone around us—family, friends, peers, and neighbors. We're also influenced by affiliation to groups, causes, and communities. In seeking acceptance, recognition, and love, you're more apt to listen to recommendations and follow someone else's lead.

The above three tiers are deeply driven by personal and immediate benefits. We start to see this shift in the remaining tiers, which are motivated by outlook and personal responsibility to preserve the planet.

- *Esteem Needs*: Once our social needs are met, we're primed to step into self-awareness. This is the moment where we evaluate our lifestyles, make habit changes, and question our impact on the environment.
- *Self-Actualization*: The top tier is saved for individuals who are willing to relinquish complacency and convenience in exchange for urgent action. Lifestyle habits and changes and purchasing decisions are influenced by the existing and future climate change impacts on our planet. In our research, 71% of "Super Greens," or the top echelon of green consumers, reported dedicating a lot of time to living a more sustainable lifestyle (Bennett and Williams 2011).

Let's put the *Hierarchy of Green* into context. Imagine if Jessica, an urban city-dwelling mother of two, needs a new bathroom cleaner. Following our assumption, Jessica would think and behave in the following way at each tier level. You'll also find questions and considerations to guide your thinking and research.

- ***Basic Needs/Are My Basic Needs Met?*** "I just need a bathroom cleaner that works. All I care about is convenience and cost."
  - Questions & Considerations: How can your product address her immediate needs? Where is she most likely to go shopping for convenience and cost?
- ***Safety & Stability Needs/Is This Good for Me and My Family?*** "I need an affordable bathroom cleaner that is *also* healthy for my family." (In this case, green benefits become a "yes, and …")
  - Questions & Considerations: How might you make Jessica feel confident in both the price, performance, and safety of your product? How might you tap into Jessica's need to protect her family? Is your greener product priced within reach when compared to other products?
- ***Social Needs/What Do My Family & Friends Think?*** "I need a bathroom cleaner, and my sister recommended this great brand that really works.

My friend also recommended another more expensive, eco-friendly brand that claims to be non-toxic. I'm thinking of getting it."

- Questions & Considerations: Who is in Jessica's ecosystem of influence and what role does she play among them? How might you leverage her influencers to tell your story? Is the performance of the product something that could be shared on social media?

- *Self-Worth Needs/Does It Make Me Feel Good?* "I need an affordable bathroom cleaner that works, is healthy for me, and is something that I can feel good about using. I am likely to research the brand's reputation, read and evaluate labels, and am more conscious of my purchasing decisions."
  - Questions & Considerations: How might your product meet Jessica's need to feel confident about her choice? How might your product tap into her growing need to be more responsible environmentally and socially? Does your product messaging make her feel good about the purchase?

- *Beyond Self-Needs/What's Best for the Planet?* "I need a bathroom cleaner that is healthy for the environment, healthy for me, and recyclable. I don't care so much about the cost; I just want the best. And if I can't find a product that fits my needs, I will make this product myself using natural ingredients."
  - Questions & Considerations: How might you use information to instill trust in Jessica? How might you leverage her passion and initiative to influence others?

Grounded in a deep understanding of human psychology, the *Ogilvy Hierarchy of Green* reminds us that we are all human at the end of the day. We are all driven by the same desires and needs—regardless of socioeconomic status, geographic location, or political party. We value clean air, clean water, and shelter. We value safety, love, and belonging, and respect for ourselves and others. We all value freedom, choice, and progress. These human drivers transcend time and technology. We just need a little nudge to get us there environmentally.

NINE WAYS TO CLOSE THE GREEN GAP

Behavior change is an opportunity to close the Green Gap. This is a gradual process in which nudging consumers toward green behaviors will shift attitudes and beliefs over time. It's a long-term play with promising return on investment (ROI). Behavior change requires inventive ways to reach, engage, and must be human-centered. With deep insights into your consumer's needs, you'll be able to craft communications and programming that convince individuals that:

- They have a personal stake in the issue or outcome.
- They can change their behavior.
- Behavior changes will result in benefits they care about.
- The benefits outweigh the costs.
- Services and products are available to help them.

In *Mainstream Green*, we used our research to identify ways we can crack one of marketing's thorniest problems: making green mainstream. Here are some of our recommended best practices to close the Green Gap (Bennett and Williams 2011).

1. Make it normal
   Everybody is doing it. Normal is sustainable. Normal drives the popularity needed for a mass movement. As marketers, our predominant instincts in the sustainability space have been to market greener products as cool or different and to confer exclusive, early-adopter status on those enlightened consumers who join in. Most of those who want to go out on a green limb and self-identify as green leaders have likely already done so, but the average consumer isn't looking for things to set themselves apart from everyone else. The average consumer wants to fit in.
   When it comes to driving mass behavior change, we need to make it normal. Even the *bona fide* greenies want to fit in more than we thought to avoid the social stigma often associated with being an environmentally conscious consumer. Ogilvy's Rory Sutherland describes it this way: "Most people, in most fields of consumption, most of the time are Satisficers. They are simply trying to avoid making a decision that is actually bad or which might cause them to look or feel foolish. The vast bulk of money in any market at any time is in the hands of Satisficers." So how can brands market their greener product and service offerings in a way that makes consumers feel normal? The first principle: make them feel like everybody's doing it.
   WeSpire calls this "**social norming**." WeSpire provides a Software as a Service (SAAS) platform to companies for motivating employees toward achieving sustainability impact goals. The platform's social networks, built on game mechanics, aim to infuse lightheartedness into employee engagement programs. By leveraging social networks inside a company, WeSpire found that employee engagement could not only catch on, but could also drive exponential growth. When a company initially starts using their platform, the likelihood that a person will take an action is about 0.2. About a year later, a person in a workplace was five times more likely to perform that same action. "Socializers are the key to normalizing behavior, and they become your most important catalyzers in the workplace," says WeSpire CEO Susan Hunt-Stevens. "It starts out linear, but then the exponential effect takes over."
2. Make it personal
   Find the "me" in green. Ask not what the consumer can do for sustainability; ask what sustainability can do for the consumer—and then show them. OgilvyEarth has a long belief that we need to shift sustainability marketing from polar bears to people. Messages that are personal resonate more deeply with people than messages that are abstract, lofty, and remote. **Companies that can link their products to highly personal benefits are better positioned to succeed**.
   This accounts for the success of certain greener product categories such as organic foods. Organic sales have doubled in the last decade. In 2022,

the organic industry grew 4% in 2022, totaling a record $67.6 billion (OTA 2023). Consumers understand the importance of organic food because it is something that they put into their bodies and is perceived to have direct personal benefits—improved quality, taste, and "purity" owing to the absence of synthetic hormones and pesticides.

3. Make green the default

Make green more convenient and widely available. If green is the default option, people don't have to make the decision. Being green in a society where green is not widely adopted is hard, even for someone deeply rooted in the cause. Being green can mean being faced with complex choices and trade-offs in what often becomes an exhausting effort to do the right thing. But what happens if you do the hard work for people? What if you make it normal by making the better choice the default?

In the United Kingdom, Marks & Spencer first introduced a 5 pence charge on food carrier bags in 2008, which successfully drove a reduction of 75% in usage. Later, in 2015, after the government made the charge mandatory, retailer Tesco reported the number of plastic bags taken home by shoppers at stores in England dropped by almost 80%.

Some credit the little English town of Modbury, with a population of 1,533, for leading the way. Modbury had the foresight to ban plastic bags in 2007. To date, 12 US cities have a plastic bag ban or charge a plastic bag fee, including Los Angeles, San Francisco, Seattle, and New York City. In 2019, lawmakers put forth 95 bills proposing a ban or fee on plastic bags. The default movement is gaining momentum (NCSL 2021).

Sometimes the best thing to do in the sustainability space is to remove the burden of complex choices from our overburdened consumers. Convenience has always sold, and making green convenient is a powerful inducement.

4. Eliminate the sustainability tax

Don't tax virtuous behavior. Governments use taxes to change behavior. Since they want fewer people to smoke, they put a hefty tax on smoking. In Russia and other low birth-rate countries, governments encourage procreation by bestowing generous tax breaks to those willing to do their part to bring children (future taxpayers) into the world.

In the greener products market, the opposite incentive is going on. We tax virtuous behavior. The premium prices of many greener products on store shelves discourage purchases and perceptions. More generous government subsidies for carbon-intensive coal and oil than for clean solar and wind energy offer the same impression.

As one of the leading providers of groceries in the United States, Walmart is in a prime position to impact the price of healthy, sustainably sourced foods. The retail giant is working to reduce the price of foods made with whole grains, as well as fresh fruits and vegetables, and they are even willing to cut into their own profits to do so (Peterson 2016). While they hope that the volume of sales will make up for the reduced profits, they stand by their belief that the customer should not have to sacrifice healthier or more sustainable options based on limits of affordability.

5. Bribe shamelessly

Incentivize progress, not perfection. Gold stars, prizes, and coupons—we all love rewards for our good behavior. Those new to this brave new world of greener choices may find themselves launched into a maelstrom of conflicting emotions, feeling they can never do enough and burdened with the curse of consciousness that comes with the first bite into the green apple. We can lighten this burden by offering them incremental, ongoing rewards for what they do accomplish, creating a framework that rewards individuals as they move along the green continuum. Since this is an imperfect journey, we are all taking together, why not make it more enjoyable, with treats along the way?

RecycleBank, for example, rewards consumers for recycling on an ongoing basis with redeemable "points" for a range of free or discounted products. When designing their rewards program, they made sure not to confuse the desired behavior (greener energy use) with the reward. RecycleBank isn't rewarding eco-friendly behavior solely with eco-friendly rewards. It is a rewarding good behavior with normal things we all want and enjoy.

This type of approach has been so successful that new entrants are rushing to attach rewards to other everyday sustainable acts such as saving energy and water. Platforms such as EarthAid, EcoBonus, and Greenopolis all partner with businesses to put money back into consumer hands for choosing more sustainable behaviors or purchasing more sustainable products.

6. Don't stop innovating

Make better stuff. We don't like going backwards. High-performing sustainable choices are key for mass adoption. Consumers are unwilling to sacrifice quality for sustainability, and rightfully so.

Unilever's Persil Small & Mighty concentrated laundry detergent saves 35 million liters of water a year in Europe and comes with a trusted brand name. Levi's successfully started a Water<Less jeans collection that now constitutes 67% of all Levi's products, saving approximately 13 billion liters of water since the program started in 2011 (Grader 2022). Companies such as Nike saw the performance challenge as an innovation opportunity. Sustainable materials for athletic shoes led to increased comfort and performance.

Thanks to a long history of premium pricing for green, the bar for more sustainable products is higher. It's not enough to perform just as well; products must perform better. In an era when opportunities to differentiate our products and brands are increasingly hard to come by, sustainability can provide fertile ground for breakthrough innovation for those marketers brave enough to turn green into gold.

7. Lose the crunch

Drop the "G" word. Green marketing needs to be more mainstream hip than off-the-grid hippie. Not everything sustainable needs to come in brown burlap and a kale smoothie green. We need to ditch the crunch factor of green and liberate ourselves from the stereotypes. And the best way to do it may be not to mention the "G" word at all.

Julie Gilhart, former fashion director for the uber-trendy Manhattan department store, Barneys NY, and a sustainability change-agent, describes how she couldn't understand why the first fabulous, eco-friendly goods she brought into the store weren't selling as well as other items. She decided to try a different approach and removed all reference to "eco-friendly" from the labels. Sure enough, sales of the premium-priced garments picked up. She realized her discerning shopper had been turned off by the crunchy image and inferior quality the eco-friendly label cued. But in its absence, the benefits of the eco-friendly materials and production process spoke for themselves: softer, more luxurious fabrics for a premium garment.

Chevy hit the right note in its Volt positioning—a high-performing car that just happens to be sustainable. Its tagline says it all: "It's more car than electric." We call this messaging approach "P.S.: We're sustainable." Communications should embrace the fact that sustainability is a dealmaker, not a deal-breaker, for the mainstream consumer.

8. Avoid making green girly
Turn eco-friendly into male ego-friendly. Green is not a sustainable proposition for the "manly man." Carry a tote, give up your truck, compost. It's true that the everyday domestic choices we need to make in favor of sustainability do not make the average NASCAR fan's heart race. Marlboro famously cracked this code when it replaced "Mild May" in its ads with the now-iconic Marlboro Man. This strong, silent type turned smoking filtered cigarettes into a manly man's thing almost overnight. Sustainability could use its Marlboro Man moment.

So how can we make green man-friendly? Let's look at a quick comparison between the Prius and BMW's eco-friendly car line, Efficient Dynamics. The Prius targets early adopters with its quirky shape, advertising humor and a focus on the environment. If you're targeting early adopters looking to telegraph their green credentials, this approach is perfect. But this campaign conveys the message that cars are bad and must be neutered. That approach will never win over more mainstream men who want their car to tell the world how manly and successful they are. BMW taps into this desire, focusing on masculine interests such as performance, innovation, and the new frontier in luxury driving to appeal to male consumers. Other brand examples include Patagonia and Clif Bar, who leverage the love of the great outdoors to inspire surfers and snowboarders to care about compromised surf and snow.

9. Make it tangible
Localize it to make it real. Sustainability is harder to follow when you can't see the trail. Find ways to help consumers see the unseeable. The line from shopping cart to the Arctic is a long one. And if the carbon footprint calculation isn't easy, even for scientists, then what should we expect from consumers at

point-of-sale? We need to simplify mental accounting and translate the murky benefits of sustainability into something immediate and concrete.

Automobiles are a major source of consumer greenhouse emissions. Most insurance companies don't consider miles driven, whether you drive 50 miles a week or 500 miles. Your car still expels carbon, which is left behind you—out of sight and out of mind.

Calculating that abstract, invisible impact isn't on the agenda for most American drivers, primarily because the consequences are far-off, indistinct, and indirect. But what if the impact of driving could be felt immediately— say, on the wallets of drivers? Currently, car insurance costs the same irrespective of mileage. What if insurance was tied to how much you drive?

In New York, Progressive has begun to experiment with a pay-as-you-drive insurance policy, while California and Massachusetts are taking the lead as part of major climate initiatives. Closing the feedback loop makes the connection between an action—driving—and its dual impacts—on the climate and your wallet—immediate and direct. The Prius came at it another way, giving drivers real-time, on-the-dashboard visibility into the impact of their driving decisions onmiles per gallon (MPG), along with bar charts for feedback over time. This was a significant contributor to the Prius' success. Tying this to dollars saved could make the tool even more motivating. Tangible signals can help consumers close the feedback loop on their purchasing decisions.

TAKE COURAGE

In the absence of courage, nothing worthwhile can be accomplished.

David Ogilvy

Courage might seem like a strange point to include, but it is so often forgotten in the malaise of back-to-back meetings, stakeholder input, and approval processes. Could it be the most overlooked step to spark change, assume a leading role, and become a memorable, meaningful brand? To put it in the words of David Ogilvy: "in the absence of courage, nothing worthwhile can be accomplished."

It can be tempting to resign ourselves to the fact that we are merely marketers, communicators, social strategists, brand storytellers. Our job is simply responding to the needs of our clients. How can we really help to drive change? Are we able to change the world for the better?

David Ogilvy wrote that advertising reflects the mores of society but doesn't influence them. History has proven him wrong. Coca-Cola's famous "Boys on a Bench" ad placed young Black and White people together, even touching, and enjoying a Coke. You've seen the ad. Now look closer. It's 1969, and these boys are sitting, integrated, on a segregation bench. That image helped normalize something that was once forbidden. Our society was becoming more—not less— diverse. This was a brave decision at the time, but a smart one based on the way the world was changing. Reversing course on inclusive advertising would have alienated the majority-in-the-making. Today, a brand could lose the trust of a whole generation.

Our work influences the mores of society, and it is our duty to our brands and our clients to do so the right way. Climate change is real. There will be those in our own companies and agencies who will advocate the "safe choice." We will hear from those who think sustainability isn't a winning message, who think that we ought to bury climate change and placate the skeptical voices in our culture. We must do the exact opposite.

In business, our view is the long one. We work for quarterly results and plan for long-term performance. We must do the same with our marketing. It's true that we cannot fail to listen to the consumer and deeply understand them. However, we must also recognize that the path toward a sustainable future is fraught with challenges. To bring about real change requires passion, determination, and most of all, courage.

NOTE

1. Our secondary research shows that context is critical and culturally nuanced. People of resource-constrained countries like India are observed to be naturally more green out of necessity. Poverty curbs consumption choices and behavior. Not to mention, poorer countries are disproportionately impacted by climate change (Eccleston 2008). Wealthier countries like the United States, on the other hand, consume unsustainably. For others, sustainable behavior is law-binding. In example, the environmental stewardship of the Iroquois of North America is both a law and a collective spiritual belief in only making decisions that benefit up to seven future generations.

REFERENCES

Bennett, G., & Williams, F. (2011). Mainstream Green: Moving sustainability from niche to normal. In *Ogilvy & Mather*. Ogilvy & Mather. https://www.ogilvy.com/sites/g/files/dhpsjz106/files/pdfdocuments/Ogilvy_MainstreamGreen.pdf. (Accessed 4/26/2024).

Eccleston, P. (2008, May 7). Poorer nations care more for the environment. *The Telegraph*. Retrieved August 26, 2023, from http://www.telegraph.co.uk/news/earth/earthnews/3341660/Poorer-nations-care-more-for-the-environment.html

Grader, K. (2022, March 21). *World Water Day 2022: Recycle and Reuse for Water Conservation – Levi Strauss & Co*. Levi Strauss & Co. Retrieved August 26, 2023, from https://www.levistrauss.com/2022/03/22/world-water-day-2022-recycle-and-reuse-for-water-conservation/

Maslow, A. H. (1943). A theory of human motivation. *Psychological Review, 50*(4), 370–396. https://doi.org/10.1037/h0054346

National Conference of State Legislatures [NCSL]. (2021, February 8). *State plastic bag legislation*. Retrieved August 26, 2023, from https://www.ncsl.org/environment-and-natural-resources/state-plastic-bag-legislation

Organic Trade Association [OTA]. (2023, May). *U.S. Organic Industry Survey 2023*. Retrieved August 26, 2023, from https://ota.com/news/press-releases/22820

Peterson, H. (2016, June 6). Walmart is about to get a lot cheaper in one key area. *Business Insider*. Retrieved August 26, 2023, from http://www.businessinsider.com/walmart-slashes-grocery-prices-2016-6

10 Aspects of Green Marketing

Al Iannuzzi

This chapter discusses key components of green marketing and sustainable brand marketing.

Greenwashing: While everyone loves the accolades and notoriety given by top management for winning some external sustainability award, all of that is so easily forgotten when your reputation is damaged because of misleading advertising. Marketers have to be cognizant of making sure that all the I's are dotted and T's are crossed when it comes to environmental claims. Reviewing TerraChoice's 7 Sins of Greenwashing will help us avoid such pitfalls.

Cause Marketing and Brand Purpose: Cause marketing is a partnership between a for-profit corporation and a nonprofit organization. These relationships are mutually beneficial; when consumers choose a brand to support a worthy cause, the firm receives a halo effect. Let's explore the approaches of leading brands like Marks & Spencer, Whole Foods Market, Tesla, Burt's Bees®, and Home Depot's approach when it comes to brand purpose and cause marketing and use of eco-labels.

GREENWASHING

With the advent of green marketing comes **greenwashing**. I was taught long ago that anything that has value will be mimicked with a counterfeit. This holds true with greener products. It's a lot easier to slap a label on a product that has a general term like eco-friendly or eco-conscious, rather than develop a truly greener product based on a strict set of criteria. Daniel Goleman, the author of *Ecological Intelligence*, states that greenwashing "pollutes the data available to consumers, gumming up marketplace efficiency by pawning off misleading information to get us to buy things that do not deliver on their promise." Additionally, he states that greenwashing "undermines consumer trust, it devalues sound data, instilling doubts and cynicism in customers" (Goleman 2009).

> **Green-wash**, the act of misleading consumers regarding the environmental practices of a company or the environmental benefits of a product or service.

During training, I always caution to never forget the basics. While everyone loves the accolades and notoriety given by top management for winning some external sustainability award, all of that is so easily forgotten when your reputation is damaged

DOI: 10.1201/9781003441939-13

because of misleading advertising. Marketers have to be cognizant of making sure that all the I's are dotted and T's are crossed when it comes to environmental claims. An increase in market share can easily be overshadowed by accusations of greenwashing in the press.

Greenwashing can range from outrageous and ridiculous to toeing the line with good intent. In my research, I came across a particularly amusing example: an image of a Hummer, a vehicle notorious for its poor fuel efficiency (some models report only 10–12 miles per gallon), covered in green logos advertising "eco-smart" and "go green" on it.

On the other hand, a challenging case arose when class-action lawsuits were filed against SC Johnson for using "GreenList" labels on its Windex and Shout cleaning products. The lawsuit asserted that the label was misleading because it gave the impression that the product had been certified by a third party, when the certification was the company's own (Vega 2010). SC Johnson had a very convincing greener product story and was one of the pioneers of green chemistry, so there was not intentional greenwashing here; nevertheless it highlights the need for transparency with sustainability claims. This case highlighted the importance of including all necessary caveats when using a logo on pack —a valuable lesson for us all.

> What we want to prevent is *Marketers Gone Wild.*

In a report by Ogilvy titled "From Greenwash to Great," they make the case that it seems that ***most greenwashing stems from marketers rushing to respond to consumer desire for greener goods and services*** and, in the process, falling prey to the overwhelming complexity of achieving corporate sustainability (Ogilvy & Mather 2010). Marketers are known for their fast-paced, competitive nature, but this enthusiasm must be tempered with team efforts to ensure factual credibility. Whenever I speak on preventing misleading claims, I often say, "What we want to prevent is **Marketers Gone Wild.**" Without the proper checks, it can be very tempting to say things that make your brand look good but are not completely true.

One of the most prominent discussions of greenwashing has come from the consulting firm TerraChoice, acquired by UL Solutions. Evaluating their Seven Sins of Greenwashing offers valuable insights on what not to do when communicating about sustainable products. I have found this useful to share with marketing teams to provide a clear understanding of what greenwashing really is.

TerraChoice's Seven Sins of Greenwashing

1. *Sin of the Hidden Trade-Off*: Suggesting a product is "green" based on a narrow set of attributes without attention to other important environmental factors. Paper, for example, is not necessarily environmentally preferable just because it comes from a sustainably harvested forest. Factors like energy consumption, greenhouse gas emissions, and water and air pollution may hold equal or greater significance.

2. *Sin of No Proof*: An environmental claim that cannot be substantiated by easily accessible supporting information or by a reliable third-party certification. A common example is when paper products claim various percentages of post-consumer recycled content without providing evidence.

3. *Sin of Vagueness*: Claims that are so poorly defined or broad that it is likely to be misunderstood by the consumer. "All-natural" is an example. Arsenic, uranium, mercury, and formaldehyde are all naturally occurring, yet poisonous. "All-natural" isn't necessarily "green."

4. *Sin of Irrelevance*: Making an environmental claim that may be truthful but is unimportant or unhelpful for consumers seeking greener products. Labeling aerosols as "CFC-free" is a common example, since it is a frequent claim despite the fact that chlorofluorocarbons (CFCs) are banned by law.

5. *Sin of Lesser of Two Evils*: Claims that may be true within the product category but distract the consumer from the larger environmental impacts of the entire category. Examples of this category include fuel-efficient sport-utility vehicles.

6. *Sin of Fibbing*: Making environmental claims that are simply false. The most common examples are appliances falsely claiming to be Energy Star certified or registered.

7. *Sin of Worshiping False Labels*: A product that, through either words or images, gives the impression of third-party endorsement where no such endorsement exists. For instance, the SC Johnson logo did not clearly indicate that the logo was for an in-house program, not a third-party certification (UL Solutions 2024).

In evaluating the Sins of Greenwashing report, the **top three mistakes made by marketers are no proof, vagueness, and worshiping false labels**. It is a fair statement to say that it isn't easy to avoid, though some marketers purposefully try to mislead customers, the majority struggle due to the newness of this type of marketing.

According to TerraChoice, one area that has resulted in significant greenwash is false labels, which are labels "associated with a product that are typically self-generated and intended to create the *appearance* of third-party endorsement." Most use the terms "eco," "environment," "environmentally friendly," or the like. To ensure that false labels are not used, it is recommended that a company uses third-party standards and certifications (UL Solutions 2024).

One of the organizations that give guidance on environmental labels is the International Organization for Standardization (ISO). This is an association of standards bodies from over 160 countries that promotes the development of voluntary, consensus-based International Standards that provide solutions to global challenges. This includes environmental labeling standards. ISO standards are maintained for five different types of labels and claims ISO14020—14024. These labels align into three types of labels (ISO 2024.

TYPES OF ECO-LABELS

Type I—"Type I environmental labelling programmes, more commonly known as ecolabelling schemes. These schemes award a mark or logo to products or services upon fulfilling a set of criteria."

Type II— "Identifies and clarifies a number of commonly used terms used in claims, whether they be on the product or elsewhere such as in product literature, advertising or reports. It also details the evaluation methods for each term in order to help ensure they are valid and scientifically sound. In addition, it has a comprehensive list of general requirements for the use of other terms that are not already defined. It also describes other label-related information and guidance, such as the use, placement and size of symbols and graphics."

Type III—"Establishes the principles and procedures for developing the data for such declarations and the requirements for declaration programs, including the requirement that data are independently verified" (ISO Environmental Labels 2024).

According to TerraChoice, companies that have their claims endorsed by certifiers that are in line with ISO14024 are safe from greenwashing (UL Solutions 2024). This ISO standard provides guidance on developing third-party labeling programs that verify the environmental attributes of a product via a seal of approval. The standards and certifications that TerraChoice feels are the most credible are listed in Table 10.1.

Business-to-Business (B2B) marketing is not exempt from greenwashing. Consider the scorecards and proposal requests that are asking for all kinds of sustainability data. I am sure there will be repercussions for misleading or erroneous information given to their business customers. The same amount of care should be taken with B2B sustainability claims as with consumer-facing sustainability claims. At a minimum, greenwashing could cause damage to the business relationship and reputation of the supplier.

Before leaving this topic, I would like to give a good example of how a credible and well-thought-out green marketing campaign can be developed. Hellmann's UK,

TABLE 10.1

TerraChoice Recommended Standards and Certifications (Iannuzzi 2018)

Terrachoice's List of Legitimate Environmental Standards and Certifications	
Biodegradable Products Institute	Natural Products Association
Chlorine Free Products Association (CFPA)	Nordic Swan
CRI Green Label	Programme for the Endorsement of Forest Certification (PEFC)
EcoCert	Rainforest Alliance
EcoLogo	Scientific Certification Systems (SCS)
ENERGY STAR	Sustainable Forestry Initiative (SFI)
OKO-TEX	Skal EKO
Fair Trade Certified	Soil Association
Forest Stewardship Council (FSC)	UL Environment Environmental Claim Validation
Green-E	UL Environment Energy Efficiency Verification
GreenGuard	USDA Organic
Green Seal	Water Sense

a division of Unilever, wanted to make their mayonnaise more sustainable. They evaluated their raw materials and determined that eggs are one of the three main ingredients. Looking at the supply chain, they made the switch to free-range eggs (Unilever 2017). This is a good example of making positive changes and appropriately communicating a sustainability improvement to customers.

WATCH YOUR CLIMATE CLAIMS

With the focus on climate initiatives there have been many statements and claims by companies to take advantage of their customers' desire to buy products from addressing this important issue. This makes it a very obvious place to be aware of greenwashing. In fact, in an analysis of 24 major global companies' climate pledges, the Corporate Climate Responsibility Monitor found that "not a single one received a 'high integrity' score and only one received a 'reasonable integrity' score" (Changing Markets 2023).

GREEN HUSHING

Not only do we have to be concerned with greenwashing but now there are accusations of **green hushing**! Green hushing as discussed by the consulting firm South Pole refers to **when a company purposely does not publicize their sustainability goals**. In a survey they conducted that included 1,200 companies with heavy GHG emissions, 23% did not plan to publicize emission reduction targets (South Pole 2022). Green hushing happens because many of these companies do not want to be scrutinized if they fail to meet a publicly communicated target and do not want to be called out for greenwashing. This is all about transparency and fear of backlash for not meeting goals. Based on my experience the best way to avoid green hushing and greenwashing is to be authentic, set goals that are reasonable, be transparent about your progress, and let everything you say publicly be backed by science and data.

REGULATORY STANDARDS FOR GREEN MARKETING

There are governmental regulatory schemes that cover environmental product-related claims, and we will evaluate three of key ones: (1) United States Federal Trade Commission (FTC) Green Guides, (2) Canadian Competition Bureau (CSA) Environmental Claims Laws, and (3) the UK's Department for Environment, Food and Rural Affairs (DEFRA). In 2023, the European Commission proposed a new law on green claims as well; at the writing of this book, it was not finalized yet. The Commission states that this law is necessary because it is difficult for consumers to make sense of the many ecolabels (230 in the EU), some of which are misleading or unreliable (European Commission 2023). There are other countries in the process of issuing greenwashing avoidance regulations too: Singapore's Green Finance Industry Taskforce is to combat greenwashing and Brazil took oversight action to address greenwashing on the Brazilian stock market (Harvard Law 2023). I anticipate that enforcement actions will increase due to the growing number of green claims.

Any company considering making green marketing claims must be fully aware and compliant with these guidelines.

EUROPEAN COMMISSION STATS

- 53% of green claims give vague, misleading or unfounded information.
- 40% of claims have no supporting evidence.
- Half of all green labels offer weak or non-existent verification.
- There are 230 sustainability labels and 100 green energy labels in the EU, with vastly different levels of transparency.

(European Commission 2023)

FTC GREEN GUIDES

The Green Guides were first issued in 1992 to give guidance to ensure that environmental marketing claims are true and substantiated. "The guidance they provide includes: 1) general principles that apply to all environmental marketing claims; 2) how consumers are likely to interpret particular claims and how marketers can substantiate these claims; and 3) how marketers can qualify their claims to avoid deceiving consumers" (FTC 2023). The guidance they provide includes:

Part 260—Guides for the Use of Environmental Marketing Claims
260.1 Purpose, Scope, and Structure of the Guides
260.2 Interpretation and Substantiation of Environmental Marketing Claims
260.3 General Principles
260.4 General Environmental Benefit Claims
260.5 Carbon Offsets
260.6 Certifications and Seals of Approval
260.7 Compostable Claims
260.8 Degradable Claims
260.9 Free-Of Claims
260.10 Non-Toxic Claims
260.11 Ozone-Safe and Ozone-Friendly Claims
260.12 Recyclable Claims
260.13 Recycled Content Claims
260.14 Refillable Claims
260.15 Renewable Energy Claims
260.16 Renewable Materials Claims
260.17 Source Reduction Claims

(FTC 2022)

The rules make deceptive acts and practices unlawful, and the FTC has brought cases against companies that have crossed the line with their green marketing claims. In April 2023, proposed changes were made to update the Guides for the first time in over a decade. The proposed updates potentially address the consumer interpretation of the terms "sustainable," "sustainability," and "organic." Moreover, the changes may provide guidance on climate-change-related claims, and energy use and efficiency (National Law Review 2023).

> ## EXAMPLE OF FTC ENFORCEMENT ACTION
>
> The FTC reached a Consent Agreement with American Plastic Lumber (APL), a company that markets plastic lumber, for making misleading claims about the recycled content of their product (FTC 2014)

Marketers should be wary about enforcement taken by the FTC. An example of this are the charges against APL, a company that markets plastic lumber, regarding misleading claims about their products' environmental attributes.

The FTC states that APL's ads and marketing materials "implied that its products—and the recycled plastics they contain—were made nearly all out of post-consumer recycled content, such as milk jugs and detergent bottles. In reality, the products contained less than 79% post-consumer content. The FTC also charged that about 8% of APL's products contained no post-consumer recycled content at all, and nearly 7% of the products were made with only 15% post-consumer content." A consent order prohibited APL from making these misleading environmental attribute claims (FTC 2014).

The FTC has also gone after companies that issue environmental certifications. An example is a letter sent to five companies that issue certifications and to the firms that use the seals on their products, indicating that they were deceptive and may not comply with the FTC guidelines. Below is an illustration of what the FTC determines to be good and bad examples of a certification seal. The good example clearly labels what the improvements in the product are and the bad example makes very general claims (Figure 10.1).

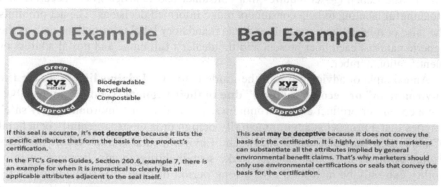

Green Certification Examples

FIGURE 10.1 Green certification examples (2015). https://www.ftc.gov/news-events/press-releases/2015/09/ftc-sends-warning-letters-about-green-certification-seals

CANADIAN COMPETITION BUREAU (CSA) ENVIRONMENTAL CLAIMS LAWS

Like the FTC Guides, Canada developed three laws for making environmental claims: the Competition Act, the Consumer Packaging and Labeling Act, and the Textile Labeling Act. These are enforced by the Competition Bureau, an independent law-enforcement agency of the Government of Canada whose mission is to protect consumers from misleading advertising.

To prevent companies from greenwashing, best practices are presented for the use of labels and self-proclaimed Type II environmental claims. These claims are usually based on a single attribute (e.g., a manufacturer's claim that a product is "biodegradable") without independent verification or certification by a third party. The guide advises that claims must be "verifiable, accurate, meaningful, and reliable if consumers are to understand the value of the **environmental information they represent (e.g., their ability to protect the environment)**."

Competition Act

The Competition Act is a federal law governing most business conduct in Canada. It contains both criminal and civil provisions aimed at preventing anti-competitive practices in the marketplace. The act addresses false or misleading information and deceptive marketing practices in promoting a product or service.

The Consumer Packaging and Labeling

The Consumer Packaging and Labeling Act requires that consumer products have accurate labeling to help consumers make informed purchasing decisions. "The act prohibits the making of false or misleading representations and sets out specifications for mandatory label information such as the product's name, net quantity, and dealer identity."

The Textile Labeling Act

The Textile Labeling Act requires that consumer textile articles bear accurate and meaningful labeling to help consumers make informed decisions. The act prohibits the false or misleading portrayals and sets mandatory label information, such as the generic name of each fiber present and the dealer's full name and postal address or identification number.

An example of advice given in the guide is **to avoid claims like "safe for the environment" or "ecological (eco)" due to their vagueness**. For example, labeling a consumer product as "environmentally friendly" or "environmentally safe" implies that a product is environmentally benign or is environmentally beneficial (Competition Bureau Canada 2021). The Environmental Claims Guide provides examples of "how to" and "not to" make sustainability claims. Because sustainability can only be measured over a long period of time, it is difficult to verify statements. However, management systems are sometimes acceptable if they can be verified. An example given by the guide makes a good point and if the concept is true with any claim a company would want to make, qualify it. Claims need to be linked to a specific achievement.

EXAMPLE

PREFERRED

"This wood comes from a forest that was certified to a sustainable forest management standard i.e., a sustainable forest management standard published by CSA, Sustainable Forestry Initiative (SFI), Forest Stewardship Council (FSC), or the Programme for the Endorsement of Forest Certification schemes (PEFC)

DISCOURAGED

This wood is sustainable".

Competition Bureau Canada 2021

ENFORCEMENT EXAMPLE

The Competition Bureau issued a fine of $130,000 to EcoSmart Spas and Dynasty Spas and ordered them to stop making misleading statements that make consumers think that their Spas were Energy Star certified when they were not. The compliance order required that advertising published in stores, and on their website, that is misleading to customers is to cease and a compliance program is to be implemented to insure adherence to the regulation (Szentesi 2011). This type of public rebuke can damage a company's reputation and sway customers away from purchasing their products.

UK's DEPARTMENT FOR ENVIRONMENT, FOOD AND RURAL AFFAIRS (DEFRA)

The UK, Green Claims Guidance was developed by the DEFRA. The Guide gives clear direction to firms that want to make claims about the green characteristics of their products. There are three attributes of claims that are the essence of good green marketing: **clear**, **accurate**, and **substantiated**. DEFRA recommends the following steps and explanations are recommended for marketers who would like to make their customers aware of the environmental benefits of their products.

Step 1: Ensure the content of the claim is relevant and reflects a genuine benefit to the environment

First, consider the full environmental impact of your product (and supply chain), service or organization. Check the claim is relevant to those environmental impacts, and/or your business and consumer interests. Ensure the claim does not focus on issues of low significance or importance. When comparing products, ensure the comparison is fair and relevant.

Step 2: Present the claim clearly and accurately

Ensure the claim is presented in a way that is accurate, clear, specific, and unambiguous and is easily understood by consumers. A claim shouldn't be easily misinterpreted or omit significant information. The scope of the claim should be clear; does it address the whole product or only one part. Do not use "vague, ambiguous words (e.g., 'environmentally friendly') or jargon that may be easily misinterpreted or confuse consumers." All imagery must be relevant to the claim and not likely to be misinterpreted.

DEFRA EXAMPLE OF A POOR PRACTICE

A product sold widely in the United Kingdom claimed that a disposable cup is marked as 'compostable'. No further information is provided. The cup will not compost in a home compost bin. An industrial composter is required, so consumers can only compost the cup if their local authority collects compostable waste for industrial composting. The claim is likely to be misleading as it does not specify the circumstances under which the product is compostable and the action the consumer needs to take. The 'compostable' claim is less likely to be misleading if any caveats regarding how and where the cup can be composted are clarified on it (DEFRA 2024).

Step 3: Ensure the claim can be substantiated

There must be data to substantiate all claims. Green claims are enforced by the Advertising Standards Authority (ASA), the UK's regulator of advertising. They monitor claims to ensure they are "legal, decent, honest and truthful according to the advertising standards codes" (DEFRA 2024).

MISLEADING CLAIMS AND ENFORCEMENT

DEFRA does not enforce the regulations for making green claims, other than for the European Eco-Label which they administer in the United Kingdom. The Consumer Protection from Unfair Trading Regulations (2024) requires all information to consumers to be fair and honest. Enforcement of this regulation is done by the Office of Fair Trading. Therefore, false or misleading claims can be taken up with this Office. An example of how the threat of action by the Office of Fair Trading resulted in significant changes in marketing practices is sustainably sourcing seafood. (Consumer Protection 2024)

SUSTAINABLE SEAFOOD EXAMPLE

Supermarkets were making various claims indicating that the fishing methods were protective of the environment and fish stocks, such as sustainably sourced, dolphin-safe/friendly, responsibly farmed, responsibly sourced, environmentally friendly farms, and protecting the marine environment. When comparing these to the DEFRA Green Claims Guidance, it is obvious that there are some very general terms used.

Major retailers including Tesco, Asda, The Co-operative, Lidl, Marks & Spencer, Sainsbury's, and Waitrose were accused of using misleading claims on seafood items like canned tuna, haddock, and cod. Under threat of enforcement through the Office of Fair Trading, ClientEarth requested that all supermarkets with misleading claims remove them as soon as possible.

As a result of this potential enforcement and the potential decline in consumer trust, retailers and seafood providers responded. In fact, the largest supermarket in the United Kingdom, Tesco, pledged that all its canned tuna would be caught by the pole-and-line method. Tesco also signed an agreement with the Sustainable Fisheries Partnership to independently review its fisheries (Tesco 2011).

CAUSE MARKETING AND BRAND PURPOSE

Cause marketing is a partnership between a for-profit corporation and a nonprofit organization. These relationships are mutually beneficial; when consumers choose a brand to support a worthy cause, the firm receives a halo effect. The nonprofit benefits by receiving income to forward their mission. Cause marketing can be an important element in advancing a brand's sustainability story.

A report by Deloitte indicated that "67% of participants were willing to pay up to 41% more for products if they were sustainable, and that inclination was higher among younger age groups" (Grundmann et al. 2022). We have seen brands respond to this by focusing on their purpose and connecting their products to causes that support the brand identity.

> 67% of participants were willing to pay up to 41% more for products if they were sustainable, and that inclination was higher among younger age groups.
>
> *Grundmann et al. 2022*

The desire to support causes with your pocketbook is not limited to the United States. Research performed by the UK-based retailer Marks & Spencer indicates that when making a purchase decision, two-thirds of their customers care about ethical choices (Goleman 2009). **Smart companies should strategically partner with worthy nonprofits as part of their green marketing program when it makes sense and connects well with their brand.**

However, caution should be taken when developing a cause marketing relationship. As noted by Carol Cone, **a company can be considered greenwashing if the cause is not perceived as authentic and a natural fit with the brand.** In effective, sustainable brand cause-marketing, there is always a direct link to the cause with the purpose of the brand. Research by Kotler and Lee determined that **full marketing benefits of a cause campaign are realized when the cause is directly related to one or more of the company's products or services** (Kotler and Lee 2004).

CAUSE MARKETING EXAMPLE

There are many good examples of sustainability cause marketing. One that stands out is Adidas combating plastic waste pollution in the oceans. Adidas partnered with Parley for the Oceans, an environmental group that focuses on protecting the sea. They partnered together turn ocean plastic into sportswear, "encouraging people to run and participate in a global movement for cleaner oceans." This is a strong example because it benefited the brand and made consumers feel good about their purchase, it was more than buying a sneaker, it was helping to protect the ocean (Adidas 2023).

BUILD YOUR BRAND WITH A CAUSE

Clorox also exemplified brand purpose by partnering with the Sierra Club when they introduced their Green Works® brand of natural cleaners. This not only gave credibility to the brand's image, but also empowered customers to support the oldest and largest grassroots environmental organization in the United States.

Another Clorox brand, Burt's Bees®, initiated a "Bring Back the Bees" campaign. The initiative was not only to raise awareness about the decreasing number of bees but also to help rebuild their habitat. This connection is an excellent fit with the brand; you can't get any closer to a cause than tying it into the brand's name (Ferraro 2016).

During a drought in Arizona, home-improvement company Home Depot initiated a program to help address water shortage. Their stores participated in a program called **Use It Wisely**, a water conservation campaign initiated by the Arizona Department of Water Resources. Water conservation aligned with products that minimize customers' water use and help address a community concern. As part of the campaign, Home Depot provided information on how customers can save water, such as sweeping the driveway rather than hosing it down and installing low-flow showerheads. This also offered an opportunity to showcase water-saving products available in stores, such as organic mulch, which reduces watering needs by 25%. Home Depot benefited by supporting this cause through connecting water conservation with their products (Kotler and Lee 2004).

Cause marketing is an important aspect of green marketing. Connecting a brand's purpose and a cause hits the sweet spot between greater customer loyalty and making the world a better place. Firms need to be cautious when initiating a cause relationship to ensure the brand is coming from a place of authenticity.

TYPES OF CAUSE MARKETING CAMPAIGNS

Transactional campaigns unlock a business donation upon point of purchase whether via actual product sale or subsequent post-purchase consumer activity.

Digital campaigns utilize online micro-sites or social media platforms to unlock business donations and/or encourage consumer donations or other online tasks.

Licensing legally permits the use of an aspect of a nonprofit brand to be used by a company in exchange for a licensing fee.

Message-focused campaigns can take many formats but focus on utilizing business resources to share a specific cause-focused message.

Events partner a cause and a company to raise money via runs, walks, celebrations, etc., or raise awareness via clean-ups, health screenings, etc.

Iannuzzi 2018

THE POWER OF PURPOSE

Businesses are taking notice of the benefits of tying causes and purpose into their brands. A Forbes study indicated that 1,099 global executives said that high-growth brands (those with 10% or more annual growth) are translating purpose into action in markedly different ways from their lower-growth peers (Deloitte 2023). Considering this, it would be wise for business leaders to make sure they connect their brand to a purpose that resonates with their target market. I see this as an extremely important issue, so much so that when I was leading the development of sustainability goals at Estée Lauder Companies, we made it a goal for every brand (25+) to be connected to a significant environmental or social purpose. We set this because we know that **brands that are connected to a purpose are the most successful.**

Swedish furniture manufacturer IKEA is a prime example of purpose built into the fabric of a company; its company founder, Ingvar Kamprad, wrote a 14-page document that lays out the purpose of the company. He states that IKEA's aim is to: "Create a better everyday life for many people by offering a wide range of well-designed, functional home-furnishing products at prices so low that as many people as possible will be able to afford them."

In the food sector, Whole Foods is another example of purpose ingrained into the business. Founder John Mackey speaks about the company he founded as one that "enriches the world by its existence and brings joy, fulfillment and a sense of meaning to all that are touched by it." In Mackey's book, "Conscious Capitalism," he discusses having a **"purpose beyond profit."** Brand purpose is integral to Whole Foods, whose consumers rely on this purpose to ensure their purchases are produced in a responsible manner and meet high sustainability standards.

Tesla further builds the case for successful companies built on a purpose-driven foundation. Elon Musk founded Tesla with a clear purpose: **"to help expedite the move from a mine and burn economy to a solar electric economy."** Tesla has revolutionized the automobile market with all major manufacturer now adding electric vehicle to their offerings (Williams 2015).

ECO-LABELS

Along with the increase in green marketing, there has been an influx of eco-labels (Figure 10.2). Some are meaningful, but others are not. Many consumers want to

FIGURE 10.2 Commonly used eco-labels by leading companies.

make greener product purchases, but they want it easy. Therefore, a logo indicating that a product has environmental benefits instils confidence in its product-sustainability attributes. However, the Ecolabel Index lists **456 ecolabels** in **199 countries** in **25 industry sectors** for marketing green products worldwide. A lot of labels can lead to a lot of confusion (Ecolabel Index 2023).

Eco-label proliferation makes it difficult for consumers, business customers, and marketers alike to know if a product has been legitimately improved. As an illustration, consider that most companies use some type of wood product for packaging or paper. Some of the labels available include the Green-e certified paper, Forest Stewardship Council, the Sustainable Forestry Initiative, the American Tree Farm System, Rain Forest Alliance, Recycled Paper symbol (chasing arrows), Printed with Soy Ink, and the Tropical Forest Foundation. Which one would you choose?

To add to the confusion, there are government-issued labels like Energy Star, WaterSense, EPA Safer Choice, USDA Organic, the EU Flower, and Canadian EcoLogo. Company greener product programs can also be confused by consumers as eco-labels, such as Home Depot's Eco Options, Staples has Eco-ID, Office Depot's GreenerOffice, P&G's Future Friendly, and Target's Made to Matter. Furthermore, there are independent companies and organizations that issue their own eco-labels or certifications, such as Underwriters Laboratories (UL), Cradle to Cradle, Green Seal, GreenGuard, and Fair Trade, to name a few. Did I confuse you yet?

According to the NGO Ecowatch, they list 13 eco-lables that they feel are significant because they've been defined by independent institutions or nonprofit organizations that have set meaningful criteria that companies must prove they've met in order to use the eco-label in question.

1. USDA Organic
2. Green Seal
3. Scientific Certifications Services
4. Forest Stewardship Council (FSC)
5. LEEDn
6. Energy Star
7. Veriaflora Responsibly Managed Petlands
8. Fair Trade Certified
9. Certified Humane
10. Leaping Bunny
11. Certified Sustainable Seafood
12. Demeter Certified Biodynamic Farm Standard
13. Non-GMO Certified

(*Ecowatch 2024*)

The complexity and amount of labels may make marketers pause pursuing an eco-logo for consumer facing products. I have seen this reluctance play out in the companies I have worked for. **Marketers do hesitate to add the cost of pursuing an eco-logo to the brand if they don't see a significant benefit**. However, the endorsement received from the right **third party can lend more credibility to a brand**, bolster its green appeal, and defend against greenwashing. According to the TerraChoice's Seven Sins of Greenwashing study, products certified by an ISO 14024-based program (e.g., EcoLogo, Green Seal, and Nordic Swan) were "more than 30% sin-free (compared to the 4.4% study-wide result). In other words, good eco-labeling helps prevent (but doesn't eliminate) greenwashing" (UL Solutions 2024).

I like to say, "choose your ecolabels wisely." What you have to ask is what will this label do for my product, will it resonate with the target market, is it worth the time and effort?

The question marketers need to ask is: Will an eco-label make a difference to my customer? If a label is desired, then which one should I pursue? One way to make this decision is to consider the labels most widely used by the leading companies we studied in this book. A review of Figure 10.2 indicates the most used eco-labels, which I noted when analyzing leading companies' green marketing programs.

Eco-labels can be helpful to demonstrate to consumers and B2B customers that environmental improvements have been built into a product. This is especially true for third-party-certified labels. However, marketers must determine if a label will help emphasize a product's greener benefit. As in most situations, it depends on what the goal of a product is. **Getting an eco-label makes sense for certain brands—but not all**. Several factors must be accounted for, including the actual market perception of a brand, and resonance of a specific label to the targeted customer segment. The use of company-generated labels can also be helpful to highlight to customers that their products go through rigorous "greening" before coming to market; however, there must be rock-solid data to back all sustainable attributes.

Having a third-party ecolabel can help with building customer trust in what your brand stands for. I have had to advise on the use of ecolabels, and I like to say, "choose your ecolabels wisely." It is not a simple thing to get a meaningful ecolabel, it takes resources, both time and money. What you have to ask is what will this label do for my product, will it resonate with the target market, is it worth the time and effort?

CONCLUSION

There are a lot of things to be concerned with when it comes to Green marketing. There is the fear of being accused of greenwashing and more recently even green hushing. What marketers must be most concerned with is providing facts and data to substantiate any claims that are made to their customers. On top of this there are regulations for how to appropriately market the greener attributes of products and new ones that are on the horizon across the world. We also should be mindful of connecting our brand to a cause and having a strong brand purpose, this resonates really well with customers and if we want to pursue an ecolabel, this should be done with much care. Green marketing isn't easy, but it can genuinely help your brand grow market share and is a strong desire of both B2C and B2B customers.

REFERENCES

Adidas X Parley. For the Oceans. https://www.adidas.com/us/parley (Accessed December 3, 2023).

Changing Markets. (2023). *Feeding Us Greenwash: An analysis of misleading claims in the food sector.* https://changingmarkets.org/report/feeding-us-greenwash-an-analysis-of-misleading-claims-in-the-food-sector/ (Accessed August 27, 2023).

Competition Bureau Canada. (2021, December 2). *Environmental claims and greenwashing.* https://ised-isde.canada.ca/site/competition-bureau-canada/en/environmental-claims-and-greenwashing#sec03-3 (Accessed August 27, 2023).

Consumer Protection from Unfair Trading Regulations 2008. https://www.legislation.gov.uk/ukdsi/2008/9780110811574/contents. (Accessed April 26, 2024).

Deloitte. Purpose a Beacon for Growth. https://www.deloitte.com/lu/en/our-thinking/insights/topics/marketing-sales/global-marketing-trends/2022/brand-purpose-as-a-competitive-advantage.html(Accessed April 27, 2024).

Department for Environment Food and Rural Affairs [DEFRA]. (2024). *Defra's Making Environmental Claims on Goods and Services.* https://www.gov.uk/government/publications/green-claims-code-making-environmental-claims/environmental-claims-on-goods-and-services. (Accessed April 27, 2024).

Ecolabel Index. https://www.ecolabelindex.com/ (Accessed December 3, 2023).

Ecowatch 2024. 13 Ecolables to Look for When Shopping. https://www.ecowatch.com/eco-friendly-labels-1881939362.html. (Accessed April 27, 2024).

European Commission. (2023). *Green claims.* https://environment.ec.europa.eu/topics/circular-economy/green-claims_en (Accessed August 27, 2023).

Federal Trade Commission [FTC]. (2014, July 28). *FTC approves final order in 'Green Marketing' case against American plastic lumber.* https://www.ftc.gov/news-events/press-releases/2014/07/ftc-approves-final-order-green-marketing-case-against-american (Accessed August 27, 2023).

Federal Trade Commission [FTC]. (2022, December 20). *16 CFR Part 260: Guides for the use of Environmental Marketing Claims (Green Guides).* https://www.ftc.gov/legal-library/browse/federal-register-notices/16-cfr-part-260-guides-use-environmental-marketing-claims-green-guides (AccessedAugust 27, 2023).

Ferraro, T. (2016, March 8). Lea Michele Burt's Bees Wildflower Initiative – Lea Michele Saving The Bees Climate Change. *Teen Vogue.* https://www.teenvogue.com/story/lea-michele-burts-bees-interview (Accessed August 27, 2023).

FTC. Green Guides. https://www.ftc.gov/news-events/topics/truth-advertising/green-guides (Accessed December 3, 2023).

Goleman, D. (2009). The New Math. In *Ecological Intelligence.* Random House.

Green Certification Examples. (2015, September 14). Federal Trade Commission. https://www.ftc.gov/news-events/press-releases/2015/09/ftc-sends-warning-letters-about-green-certification-seals (Accessed August 27, 2023).

Grundmann, G., Klein, F., & Josten, F. (2022). Sustainability in business: Staying ahead of the curve. *Deloitte Insights.* https://www2.deloitte.com/us/en/insights/topics/strategy/sustainability-in-business-staying-ahead-of-the-curve.html

Harvard Law. (2023, July 24). *Greenwashing: Navigating the risk.* The Harvard Law School Forum on Corporate Governance. https://corpgov.law.harvard.edu/2023/07/24/greenwashing-navigating-the-risk (Accessed August 27, 2023).

Iannuzzi. (2018). Greener Products; The Making and Marketing of Sustainable Brands. CRC Press. Terrachoice Recommended Standards and Certifications. p. 219.

Iannuzzi. (2018). Greener Products; The Making and Marketing of Sustainable Brands. CRC Press. Types of Cause Marketing Campaigns. p. 227.

International Organization for Standardization (ISO) Environmental Labels. https://www.iso.org/files/live/sites/isoorg/files/store/en/PUB100323.pdf. (Accessed April 26, 2024).

Kotler, P., & Lee, N. (2004). Best of Breed. *Stanford Social Innovation Review, 1,* 14–23. https://ssir.org/articles/entry/best_of_breed

National Law Review. (2023, June 4). Going 'green'—what Does that Mean? FTC Proposes Revisions to Green Guides. *The National Law Review.* https://www.natlawreview.com/article/going-green-what-does-mean-ftc-proposes-revisions-to-green-guides

Ogilvy & Mather. (2010, June 22). From Greenwash to Great. *PR Newswire.* https://www.prnewswire.com/news-releases/from-greenwash-to-great-91834209.html (Accessed August 27, 2023).

South Pole. (2022). *Net Zero and Beyond – South Pole's 2022 net zero report.* https://www.southpole.com/publications/net-zero-and-beyond (Accessed August 27, 2023).

Szentesi, S. (2011, January 18). Spa Retailers Out of Hot Water after Settlement with Bureau for Allegedly False Energy Savings Claims I CANADIAN COMPETITION LAW. *COMPETITION LAW.* https://www.ipvancouverblog.com/2011/01/spa-retailers-out-of-hot-water-after-settlement-with-bureau-for-allegedly-false-energy-savings-claims/ (Accessed August 27, 2023).

Tesco. (2011, January 11). *Tesco makes sustainable tuna pledge* [Press release]. https://www.tescoplc.com/tesco-makes-sustainable-tuna-pledge/ (Accessed August 27, 2023)

UL Solutions *Sins of Greenwashing.* UL Solutions. https://www.ul.com/insights/sins-greenwashing. (Accessed April 26, 2024)

Unilever. (2017, January 23). Hellmann's® Mayonnaise and Mayonnaise Dressings Now Use 100% Cage-Free Eggs in the U.S.*, Three Years Ahead of Schedule. *PR Newswire.* https://www.prnewswire.com/news-releases/hellmanns-mayonnaise-and-mayonnaise-dressings-now-use-100-cage-free-eggs-in-the-us-three-years-ahead-of-schedule-300394645.html (Accessed August 27, 2023).

Vega, T. (2010, October 7). F.T.C. Proposes Tighter Rules for Green Claims. *The New York Times.* https://www.nytimes.com/2010/10/07/business/energy-environment/07green.html (Accessed August 27, 2023).

Williams, Fryea. (2015). *Green Giants.* American Management Association.

11 Best Practices and Conclusions

Al Iannuzzi

This final chapter summarizes insights from evaluating trends and leading practices in greener product development and marketing. Interest in greener products is on the rise as consumers seek ethical, environmentally responsible businesses. From analyzing over 40 case studies, it is easy to see the best practices for making and marketing sustainable brands that make leading companies successful. Both customers and companies can make a positive impact by what they purchase and sell. Fewer resources can be used, good causes are supported, and costs are reduced. When you have both a truly greener product that is communicated effectively, *everyone wins*. Customers' needs are met, brand loyalty is built, and success is achieved. The focus on greener products and sustainable brands is here to stay and is an imperative for companies. The demand for eco-innovative products continues to become the expectation. Companies that can provide these products while minimizing costs will emerge as the big winners.

THE SUSTAINABLE BRANDS IMPERATIVE

Interest in greener products is on the rise. We have seen, through numerous studies, that customers want products that are environmentally friendly, and they want to purchase from businesses that operate in an ethical manner. According to a study conducted by IBM that included more than 16,000 respondents in more than ten countries, 77% of consumers desire to make more sustainable product choices at home (IBM 2022).

Why are consumers interested in purchasing greener products? Both Business-to-Consumer (B2C) and Business-to-Business (B2B) consumers are responding to the global green explosion. Heightened awareness of global environmental issues like climate change, scarcity of water, air pollution, and the use of toxic chemicals drives the focus on using the power of purchase to do good. Companies are responding by meeting the customers desire with their sustainability programs. Recent research indicates that the highest demand for greener products comes from countries with the highest growth potential and the most pressing environmental problems, such as Brazil, China, and India. Perhaps this is due to a heightened awareness of local issues and the subsequent desire to purchase products that will make a difference.

I believe Walmart, the consumer company with the largest sales in the world, has been the most influential in driving the development of more sustainable products, requiring suppliers to complete scorecards measuring their sustainability performance, and seeking greener products for their stores. Other major companies have followed

DOI: 10.1201/9781003441939-14

suit: Kaiser Permanente, a large US-based hospital chain, developed a sustainability scorecard for their suppliers; Procter & Gamble did the same. Retailers, such as Tesco and Marks and Spencer, and home improvement and building supply chains Lowe's and Home Depot have also emphasized offering greener products to their customers and have set aggressive sustainability goals which impact companies selling in their stores.

These initiatives have influenced product development processes and are driving competition based on sustainability. The desire for greener products touches all types of businesses and has made green marketing an imperative in B2B sales. We have seen chemical companies trying to help their customers develop more sustainable products, food industry providers focusing on fair trade and sustainably sourced raw materials, building product companies having offerings that can help customers achieve green building certifications, and apparel companies greening their supply chain.

Product ratings further the necessity to build sustainability into product development process. The Environmental Working Group and Yuka have evaluated tens of thousands of products and make it easy for the growing number of ecologically conscious consumers to choose the most wholesome product for themselves and their family. Socially responsible investment rating systems are considering the impact of products as well.

Regulatory drivers have also caused a shift in product development with the advent of environmental legislation for products. It started in the European Union with requirements for packaging design and take-back programs for electronic products. Then, it moved into restrictions of certain toxic metals, flame retardants, and other chemicals. There have been many new regulatory areas with significant impact on the way business is conducted: EU REACH, RoHS, Packaging, and WEEE Directives, etc. These product-based regulations have expanded into all regions of the world with the advent of extended producer responsibility regulations, banning of chemicals like BPA in Canada, California's Proposition 65, China REACH, Korea RoHS, Brazil packaging regulations, and the UN Globally Harmonized system for the Classification and Labeling of Chemicals (GHS), to name a few. Recent developments in Europe and even in the state of California in the United States are requiring companies to report their greenhouse gas (GHG) emissions and on sustainability initiatives like their impact on the economy, environment, nature, and communities.

It seems that the regulations keep coming and expanding into other areas like greenhouse gas reporting, modern slavery and deforestation, alongside non-regulatory pressures from NGOs against specific chemicals or compounds. We have seen effective campaigns against PVC, bisphenol-A (BPA), triclosan, phthalates, DEHP, etc. Some are based on science, and some are not so scientifically based. An important issue that companies need to be aware of is that perception is reality and that **Risk = Hazard + Outrage**. Therefore, monitoring emerging issues becomes a more critical part of product stewardship and sustainability programs. Identifying these issues and trying to shape them with sound science and dialog is essential to the mitigation of risk. These kinds of pressures will only grow in the coming years.

These are the factors driving the imperative of making and marketing greener products. So, what have we learned about how leading companies respond to these pressures and customer demands? **Robust greener product development programs address customer, regulatory, and stakeholder demands.**

BEST PRACTICES FOR MAKING GREENER PRODUCTS

Based on a review of leading companies, there are some common aspects to their greener product development programs. The most effective programs have **top management endorsement** and are viewed as a **business initiative** rather than an environmental program. A case in point is Unilever's Sustainable Living Plan; it is clearly a company priority and has boasted many accomplishments over a ten-year span, such as a 32% reduction in the total waste footprint per consumer product use (Unilever 2020). The business value of their program lies in connecting to customers' sustainability needs. Samsung's CEO oversees their Green Management Committee, marketing more energy-efficient products such as televisions and other electronics. Other companies have built "green" right into the identity of their brand, such as Seventh Generation and Method.

Having **third-party involvement** in the **development and endorsement** of eco-innovation programs is also a key best practice. GE developed Ecomagination with the help of the consulting firm GreenOrder, and they have an independent advisory board that includes academics and NGOs. BASF's eco-efficiency approach and tools are verified through two independent third-party organizations, TÜVs (German technical inspection and certification organizations) and by NSF (National Sanitation Foundation). Method uses the Cradle-to-Cradle certification and has ingredients reviewed by a third party.

The use of **scorecards, focus areas**, and **tools** to help identify the most important **product life-cycle steps** to improve upon is an important part of developing sustainable products. Examples include the Greenlist ™ environmental rating system used by SC Johnson which helps developers see a clear pathway to improving a product's environmental performance. Philips has six Green **Focal Areas to** make products greener: energy efficiency, packaging, toxic materials, weight, recycling & disposal, and lifetime reliability.

Samsung products are rated through the "Eco Design System (EDS)" and issue an Eco Rating. Products are put into three categories based on their eco-grading scheme: (1) Eco-Product, (2) Good Eco-Product, and (3) Premium Eco-Product. Method has five key design elements: (1) Clean—effective formulas that work. (2) Safe—people and pet friendly. (3) Green—safe and sustainable materials that are manufactured responsibly. (4) Design—attractive product designs. (5) Fragrance—use of flowers, fruits, and herbs for product scents.

In addition to greener systems and processes, leading companies have **enterprise-wide product stewardship goals**. These goals are important to help the entire organization rally behind and drive more sustainable product design. Clorox has Eco Goals to generate one-third of growth from environmental sustainability initiatives and make sustainability improvements to 25% of their product portfolio. Philips has EcoVision, which requires improving the energy efficiency of Philips products by 50%, doubling global product collection, and recycling and incorporation of recycled materials in products. Seventh Generation has set a goal to have all their paper packaging to contain 100% post consumer recycled (PCR). Unilever's has several objectives: cut in half their environmental footprint of the making and use of their products, help more than a billion people take action to improve their health and well-being, and source 100% of agricultural raw materials sustainably. Revenue from greener products is also part of some enterprise-wide goals; P&G has a $50 billion cumulative sales target for sustainable innovation products. Johnson & Johnson set a target that 20% of revenue from Earthwards® recognized products.

Life-cycle analysis or life cycle thinking targets improvements in the most impactful areas. Philips identified the key life-cycle aspects for each of their product categories: reducing energy consumption, weight, and dose for healthcare, energy efficiency and closing material loops (e.g., increasing material recycling) for consumer lifestyle and energy efficiency for lighting. Method considers life cycle impacts through their Cradle-to-Cradle approach, while P&G looks at a product's full life cycle (raw materials, manufacturing, and product use) to help identify the most important areas to focus on. Unilever used life-cycle assessment to identify that hot-water consumption during the use of soaps and shampoos is the biggest impact area, leading to their Turn off the Tap campaign which encourages customers to turn off the shower while they lather.

Transparency of the materials used in products has been a practice that several companies have committed to. This helps gain public trust and reinforce the credibility of the company. **People trust companies that voluntarily provide more information to them than is required**. Clorox is listing all ingredients on product labels for the Green Works® line of naturally derived cleaner. Similarly, SC Johnson has made a commitment to make their ingredients available. Both Method and Seventh Generation have transparency commitments.

Meeting customers' needs is another key characteristic of a leader in product stewardship. Providing **end-of-life solutions** for your products is an important way to address customer requirements. This is especially true with electronic products. Apple and Samsung have established take-back programs to facilitate product recycling.

BEST PRACTICES FOR MAKING GREENER PRODUCTS

- Top management support and greener products are part of the business strategy.
- Third-party input in developing design criteria.
- Use of scorecards, focal areas, and tools to make it easier for product developers.
- Enterprise-wide goals to augment individual brand improvements.
- Use of life-cycle analysis or life cycle thinking to focus on the most important impact areas.
- Transparency of ingredients used in products to build more trust.
- Meet customer requirements by providing end-of-life solutions for products.

BEST PRACTICES FOR GREEN MARKETING

The **first step to effective green marketing is to have a truly greener product**. Companies may fear making green claims, fearing backlash due to imperfections. However, I always emphasize that **there is no such thing as a green product**. The only true green product is the one you didn't use. Every product has an environmental impact, from raw materials, to manufacturing and transportation, to disposal. The key is to focus on **greener, more sustainable products**, and then continuously improve and reduce environmental impacts. Customers understand that companies are not perfect, but they do value efforts that move in the right direction. The three keys to green marketing that I have distilled are:

(1) Having a credible greener product story.
(2) Meet your customers greener product demands.
(3) Appropriately communicate the product's greener attributes.

Once you have a greener product, there are some common marketing approaches being taken by leading companies. One of the key components to successful green marketing is having an **effective communication method** to clarify how your product is meeting the customers' needs. Timberland uses a "nutritional label" that indicates how the product fares in the areas of climate impact (greenhouse gas emissions), chemicals used (presence of hazardous substances), and resource consumption (use of recycled, organic, or renewable materials). Ecomagination is the vehicle of communication for GE; products that have the Ecomagination designation provide the actual reasons why it is more sustainable, for example, less energy or water use.

Earthkeepers® is a branded way for Timberland to communicate the greener products that they offer. On their website, consumers can see the percentage of recycled content, organic materials, even information about the tannery process for Earthkeepers® products. Several other companies also use their **greener product development focus areas** to communicate to customers. Examples include the SC Johnson GreenList® and Johnson & Johnson's Earthwards®.

Another effective method of communication is product environmental profiles. Apple maintains Product Environmental Reports for their devices. The report informs customers of the impact in several important areas: climate change, energy efficiency, material efficiency, packaging, restricted substances, and recycling. This provides producers with a concise way to communicate product improvements as well as demonstrate their commitment to environmental protection.

Leading companies often use **eco-labels** to add validity to their green claims and make it easy for customers to see that the product is greener. However, with over 450 eco-labels available, only a few are recognized by consumers. The eco-labels that were most used by the leading companies studied included Recycling Symbol, Cradle to Cradle, Energy Star, Forest Stewardship Council (FSC), Sustainable Forestry Initiative (SFI), WaterSense, USDA-Certified Organic, and

Fair Trade. So if you think that an eco-label can help your brand, choose wisely. Honest Tea's use of USDA Organic, Apple's use of Energy Star, and WaterSense-labeled toilets and bathroom faucets sold at Lowe's are illustrations of effectively using eco-labels.

Leading firms used **company-branded greener product lines**, either by improving products incrementally or developing new greener products from the foundation up. In both cases, it is helpful to communicate sustainability characteristics that have been enhanced. Some of these programs have taken on eco-label-like status. Examples of using branded programs for products greened from the ground up are Clorox Green Works® Johnson & Johnson's Earthwards® and Timberlands Earthkeepers® line of footwear. These programs resonate well with B2C and B2B customers because they highlight improvements that are meaningful to them, such as energy efficiency, reduced weight, more recyclable packaging, or less hazardous substances.

Cause marketing and brand purpose are important considerations for a green marketing program. Even when the United States was in a prolonged recession, nearly three out of four Americans (72%) said that they were more likely to give their business to a company that has fair prices and supports good causes than to a company that provides deep discounts but does not contribute to good causes (Edelman 2010). Doing good through purchasing products spans the globe—**customers care about making ethical choices**.

There are numerous examples of using purpose and cause marketing to enhance a brand, such as. the Sierra Club and the Green Works collaboration. When Clorox wanted to bring greener products mainstream, they pursued added endorsement to their natural-product claims from the Sierra Club. Care must be taken when entering a cause relationship; it must be authentic and not perceived as "buying" the greener-product credentials. For instance, Haagen-Dazs ice cream relies on bees for their ingredients, and their cause is to support research to save the honeybees due to population decline, a really great connection to the brand and a winning cause connection.

The final best practice for green marketing is defensive: **preventing greenwashing**. Greenwashing accusations can severely damage a brand's reputation and make it hard to recover. To avoid this, it makes sense to put processes in place to avoid greenwash and ensure that all claims are authentic and factual. Some companies have relied on third-party eco-labels such as Cradle to Cradle, Energy Star, Safer Choice, and WaterSense. Relationships with well-known NGOs also help to mitigate false or misleading claims, since it adds credibility.

The other aspect of preventing greenwashing is to ensure that the sustainability claims do not overshadow the essence of the brand. **Sustainability claims should complement the key aspects of the brand**. One of the most significant greener products—Coldwater Tide laundry detergent—leads with the efficacy of the product and *then* mentions energy savings as a secondary message.

- Offers brilliant clean and color protections.
- Offers 50% more energy savings when switching loads from warm to cold.

Notice that the greener benefits complement the most important product feature, cleaning power (Protector & Gamble 2023).

> **GREEN MARKETING BEST PRACTICES**
>
> - Communicate greener characteristics effectively to meet customers' needs through reporting environmental benefits using product stewardship focus areas and product profiles.
> - Use well-known, respected eco-labels or third parties to endorse products.
> - Implement company-branded greener product lines and internal designations.
> - Tap into the brand's purpose and leverage appropriate cause-marketing relationships that have a direct nexus with the brand.
> - Prevent greenwashing by being authentic and not overstating greener product attributes.

CONCLUSIONS

The world's need for greener products is increasingly urgent given the increasing stressors on the earth's systems and the expanding global middle class. There is an increased concern about resources that are being demanded for products by the growing global middle class, the number of pollutants that are found in our food and even in our bloodstreams, expediting the global demand for natural or organic products.

Natural products like Method and Seventh Generation, once were considered niche, are now sought after in mainstream supermarkets. The old stereotype of green products having sub-par performance are long gone, particularly with game-changers like Green Works®, proving that large multi-nationals can develop and win in the marketplace with a naturals-based product platform.

The pull for greener products is beyond consumer marketing. B2B marketing of greener products is picking up pace. Chemical companies are making it a point to differentiate their products based on eco-innovation, indicating significant shifts. Life-cycle assessments reveal that environmental impacts often come from the acquisition of raw materials, product use, or its end of life. This knowledge is causing a shift toward greening up products in non-traditional ways. Over recent years, sustainable innovations included cold water laundry detergents, bio-based plastics, take-back programs, and sustainably sourced raw materials, to name a few.

> When you hit this *sweet spot*, of having a truly greener product that is communicated in an appropriate way, *everyone wins*.

Making products *greener* is becoming an expectation, alongside efficacy and quality. Reduced environmental impact is being viewed as an "**and**." While there are exceptions, greener products will not command a higher price. Customers want to have a high-quality product that works **and** is greener, without a premium price. This customer requirement spans all types of businesses: apparel, chemicals, building products, paper products, food, medical equipment, and packaging.

Building an eco-innovative product requires a team effort and signals from the field must be gathered to gauge customers' needs. R&D, procurement, operations, and product stewardship groups need to collaborate to build in the desired attributes. A balanced, clear communications program must be devised by marketing and delivered by sales groups. Care must be taken when communicating about greener products. First, there must be an authentic, science-based story. Second, the message must be simple and transparent to demonstrate how the product or service meets customers' sustainability needs.

Both customers and companies can make a difference by what they purchase and sell. Fewer resources can be used, good causes are supported, and costs are reduced. When you hit this **sweet spot** of having a truly greener product that is communicated in an appropriate way, **everyone wins**. Customer's needs are met, and brand loyalty is built. The focus on greener products and sustainable brands is here to stay. We can expect a steady demand for eco-innovative products in the coming years. Companies that provide these products without increasing costs will be the big winners.

REFERENCES

Edelman. Role of "Citizen Consumer" to Tackle Social Issues Rises, as Expectation of Government to Lead Declines. *PRNewswire*. https://www.prnewswire.com/news-releases/role-of-citizen-consumer-to-tackle-social-issues-rises-as-expectation-of-government-to-lead-declines-106678903.html (Accessed 2010, November 4).

IBM Institute for Business Value [IBM]. (2022). Balancing sustainability and profitability. In *IBM*. Retrieved August 26, 2023, from https://www.ibm.com/downloads/cas/5NGR8ZW2

Procter & Gamble. (2023). *Tide Plus Coldwater Clean Liquid Laundry Detergent*. Tide. https://tide.com/en-us/shop/type/liquid/tide-coldwater-clean-liquid (Accessed August 26, 2023).

Unilever. (2020, May 6). Unilever celebrates 10 years of the Sustainable Living Plan. *Unilever*. https://www.unilever.com/news/press-and-media/press-releases/2020/unilever-celebrates-10-years-of-the-sustainable-living-plan/ (Accessed August 26, 2023).

Index

Note: Page numbers in **bold** and *italics* refer to tables and figures, respectively.

Printed in the United States
by Baker & Taylor Publisher Services

Printed in the United States
by Baker & Taylor Publisher Services